Carp farming

Carp farming

V K Michaels

Fishing News Books Ltd
Farnham · Surrey · England

British Library CIP Data

Michaels, V.K.
Carp farming.
1. Carp farming
I. Title
639.3′752

ISBN 0 8528 153 0

Published by
Fishing News Books Ltd
1 Long Garden Walk
Farnham, Surrey, England

Typeset by
Mathematical Composition Setters Ltd
Salisbury, Wiltshire

Printed in Great Britain by
Henry Ling Ltd
The Dorset Press, Dorchester

Contents

Illustrations

Fig

Preface

It is mid-January here at Newhay and bitterly cold. It is snowing and all our ponds are frozen over. Beneath the snow covered ice the carp are hibernating.

In these conditions there is little to do on the carp farm except the routine walk round the ponds every day. Indoors there is always work, mostly repairing tools, nets and aerators; but not for me today. Friends are expected in the afternoon; we are celebrating my 70th birthday and with that, sadly, the day has come to retire from active carp farming.

For many years I wanted to write a book on carp, but other commitments always prevented this. Other works on carp are available now, but because of my lifetime's practical experience my wish was to produce a book which explained carp farming in plain language that the layman can easily understand.

In January 1974 I published a booklet 'First Steps in Carp Farming' and 5,000 copies were sold by 1982. In February 1983 this booklet was updated under the same title by my colleagues Keith Colwell and Martin Jaffa. To date almost 10,000 copies have been sold, both at home and abroad.

The reader may question the need of a further book, but my experience in carp farming started about 60 years ago and the booklet does not incorporate this vast practical experience which I feel should be recorded, as it appears to be somewhat unique in the UK.

I was born in Eastern Europe and as a young boy I spent most of my free time on the carp farm of a family friend. They had a son of my age and together we enjoyed playing around the 45–50 ponds of various sizes. All the ponds were situated on both sides of a stream with a water wheel at the top end, whose main purpose was to provide power to a flour mill. On the lower end another water wheel was used for cutting timber. We spent may happy hours just watching the thousands of young carp feeding on zooplankton, large carp jumping about in the water on warm summer evenings and breeding fish spawning amongst the water plants. At the same time we were aware of workmen building new ponds, catch-pits and monks and sometimes we took an active part in helping the men to set traps, pull nets or sort out carp for the market. Occasionally we were rewarded by permission to fish ourselves with rod and line. We were always disappointed if we failed to land a large fish (anglers will surely agree that it is a fine art to hook a large carp).

These were happy days which continued right up to the time when, with a certain amount of sadness, I had to leave home to study engineering. Almost immediately after completing my studies the Second World War started and with it came a chain of events which were to change my life drastically; but that is another story and will possibly be the subject of a book in the future. Perhaps just to mention here that my friend from the carp farm was killed in the war and, on a happier note, I got married.

We were living in a camp near Hanover when one day we received a visit from a delegation who were looking for volunteers to work in England. Shortly afterwards my wife and I found ourselves settled in Halifax, Yorkshire to begin a new life. We worked together in a textile mill – my wife as a twister and I as a bobbin setter. After three months I was offered a post as a second engineer in the same mill and after a further six months I became first engineer due to the previous holder emigrating to Australia. I was now in charge of a 750 hp Tandem Steam Engine, a complete mechanical workshop including tools, lathes and welding equipment. As there was an abundance of scrap angle-iron lying about, it was not long before I obtained the permission to use this and in my lunch break I began to make my first aquarium frame. The memory of the happy years with my late friend as boys on the carp farm never left me and as at that time carp farming was an impossible dream, I saw the opportunity to enjoy nature in the home.

I was still interested in the carp species which at that time were very difficult to obtain, but I managed to locate some, and I reared successfully in my first three feet tank. I continued my lunch time activities and before long I had a dozen tanks lining the walls of my home. Many of my workmates began to show an interest in what I was doing and eventually we formed an aquarium club. We met weekly in each others homes and the club flourished. Very soon we found ourselves with a surplus of our home bred fish and this triggered off my first thoughts of opening an aquarium shop in the town. The greatest doubt I had was to convince my wife that this was the right thing to do and to my great surprise and delight she agreed immediately and about six months later 'Michaels' Aquarium' was opened. The venture was a tremendous success and within 12–18 months we were both importing and exporting fish to many countries; I was the first person to import the firemouth cichlid and black shark into this country and together with two other dealers, the first to import koi direct from Japan. I kept some of these koi for many years, and I know that some of the early imports of koi are still alive today.

In the meantime the aquarium business was going very well. Better equipment became available but there was one area in which I could see room for improvement – and this was in equipment for filtering the water. The only filter that could be purchased at that time was a celluloid corner filter, which contained charcoal in the base and a piece of glass wool on the top. In order to improve on this filter I decided to use all the gravel in the aquarium as the filter medium. Using some empty baked-bean tins and the soldering iron I made various prototypes until I perfected a filter and 'Michaels' Filter-Aerator' was produced and patented in a number of countries. It was a great success financially and enabled us to achieve yet another ambition – to enter the motor trade. In 1954 we bought a garage in Selby, Yorkshire. Years of hard work followed but I never lost interest in my favourite fish – the carp.

As part of the business we had a cafe and we soon had three large six feet aquariums fitted into the cafe. Two large aquariums were built over the petrol pumps. We also had two good sized garden ponds at home. As a founder of the Selby and District Aquarium Society I became its chairman. The years went by and I derived a great deal of pleasure from keeping carp. The garage business prospered and grew from humble beginnings to a business which employed 65 staff. Unfortunately in 1969 I became ill, possibly due to overwork and my war injury. I was in and out of hospital and eventually

Fig 1 Michaels' filter aerator

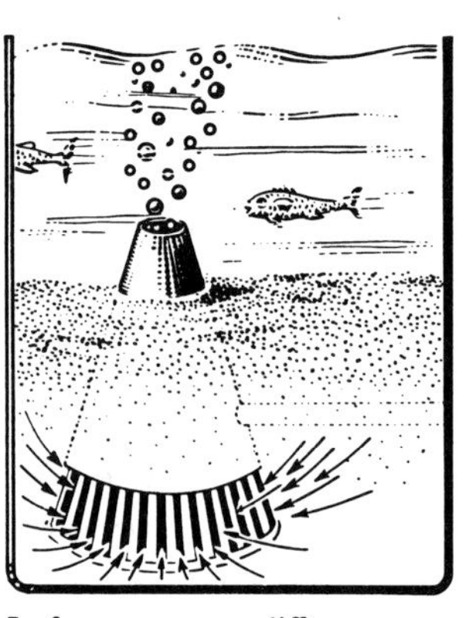

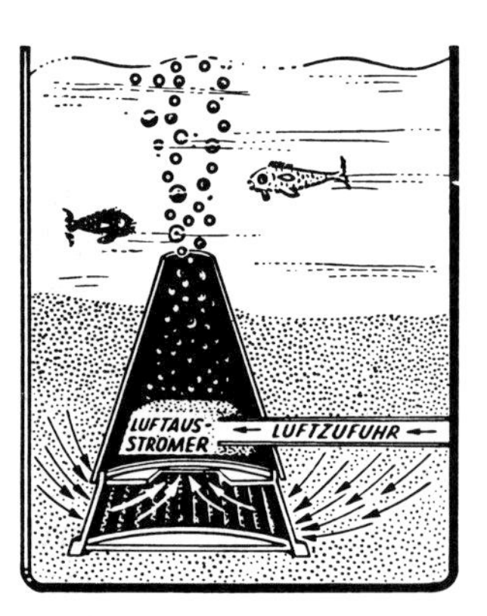

Luftaus-stromer = diffuser
Luftzufur = compressed air

Fig 2 Aquarium tanks at author's garage

on medical advice I sold the garage business and retired.

After a good rest I thought it was right to spend more time with my carp. My daughter was grown up now and the indoor swimming pool, which had been built mainly for her use, was standing empty. It could be heated and I saw that I could use it for breeding carp. Each day I took a long walk along the nearby River Ouse to exercise both myself and my dog. On these walks I passed an area of land of approximately 10 acres which was totally overgrown with reeds and willows and partly flooded. Being curious about this particular piece of land I was told that in 1947 a high tide together with heavy rainfall had burst the bank of the river there, and a new bank had to be built. In order to do this the River Authority purchased the land adjacent to the river. In 1973 we were given permission to dig carp ponds on this land and a little later we were granted permission to build a house and laboratory adjacent to it. Life for me had gone a full circle and I could once again lay in the grass watching the carp, which I had learned to respect and to love, and this time they were my own.

It has given my family and myself 18 happy years and in that time we have bred and grown hundreds of thousands of carp.

I had to tell this story and I hope that the reader found it interesting.

V K Michaels
1988

Acknowledgements

I would like to pay tribute to my former colleagues Keith Colwell, Martin Jaffa and Jeremy Clarke. Together, I feel that we have made a contribution to the carp farming industry in the UK.

I would like to thank Phyllis Clegg and Trevor Goring for their respective contributions to this book.

1 History of carp culture

There are many species of carp. Together they form the largest family of freshwater fish in the world – the cyprinids. They are found in almost all areas of the world but do not naturally occur in the polar regions, South America or Australia. Almost all species of carp have scales but none have teeth in the jaws; instead they have curved bones in the throat which are used for grinding some of their food. Most carp species are small; some, even fully grown, are not much larger than one inch in size (*eg* the attractively coloured *Rasbora maculata*). Others, like the tiger barb (*Barbus tetrazona*) and the harlequin fish (*Rasbora heteromorpha*) are even more colourful. Many of these smaller species are well known by aquarists all over the world. They thrive in captivity and have given much pleasure to millions of people. By contrast, in some tropical regions, members of the same family grow very large – the giant mahaseer from India reaches a length of two metres. In more moderate climates crucian carp, goldfish, tench and bitterling are characteristic members of this very large family. The carp featured in this book is known by the latin name *Cyprinus carpio* L (L stands for the name of the Swedish naturalist Linné 1707–1778). It is generally accepted that this species originated in the region of the Caspian Sea; from there it is believed to have been brought west by Roman soldiers – to the Black Sea and the River Danube. The

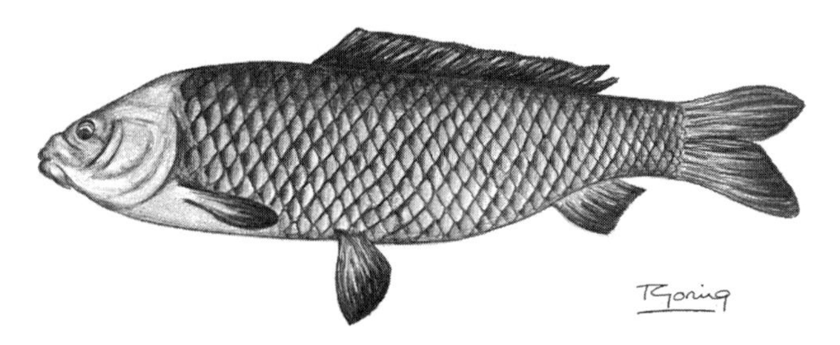

Fig 3 Original wild carp (artist's impression)

common carp also spread eastwards to China – probably introduced by soldiers of the Chinese imperial army.

The Romans, who were already knowledgeable in cultivating fish as food, found out that carp is very adaptable, grows very fast, feeds on almost anything and can exist in relatively poor water conditions. Some of these cultivated carp probably escaped into rivers, particularly the Danube, travelling west and north, adapting themselves to the climatic conditions. As time went by the carp attracted the attention of monks, who could see the versatility of this particular fish and from these beginnings carp farming slowly spread over most European countries right up to the Baltic and eventually to England.

The demand for fresh fish was growing all the time particularly in the countries which had no direct access to the sea. Practices needed to be improved to speed up the growth of carp and the efficiency of pond construction. Once suitable sites around the monasteries had been found, digging started and hundreds of these ponds, particularly on the continent of Europe, are still in production.

The ponds were constructed so that they could be drained and each had a catch-pit for easy harvesting. The sluices for draining were at that time made from timber and as this type of sluice was invented by monks it took its name from them and is still known today as a 'monk' (see section on pond construction).

Kings and Kaisers all over Europe encouraged carp

farming and in some instances actually ordered it.

Ponds had to be dug in the middle of the villages so that the water could be available for fire fighting purposes if necessary. At the same time they were stocked with carp. Many smallholders and farmers began raising the fish and very soon it became the most common table fish in the whole of Europe. As the carp is basically a warm water fish breeding was something of a problem. That was until the invention of the Dubish pond (Thomas Dubish 1813–1888 – see *Carp breeding*).

The development of carp farming on the European continent except England, took place as follows:

300–600 A.D. introduction of carp into Europe. Naturally breeding stocks established.

700–1300 A.D. many more carp brought into Europe. Successful pond culture and breeding of carp mainly by monks.

1400–1600 A.D. selective breeding began and various types of carp were developed by breeders who tried to produce ever faster growing fish. From the long, slim, wild carp much larger, but still fully-scaled carp were produced. This development was relatively easy to achieve because the carp now grown in ponds had access to a plentiful supply of better food with minimum expenditure of energy, whereas wild fish in rivers had to use more energy to survive. Consequently the farmed carp became much plumper.

Selective breeding (the fastest growing carp being used for reproduction) continued this improvement. The housewife also had some influence in the quality of carp mainly because scaly fish are difficult to clean. So again, by selective breeding, a fish emerged which we now know as a mirror carp. The scales had almost disappeared except for one row on the top of the back and some near the tail and gills. Some of the scales grow very large and are shiny – reflecting light – hence the name mirror carp. Another related variety is the zeil carp. In this fish one more line of scales grows through the middle of the fish. Another variety – the leather carp – is completely devoid of all the scales. From these four forms of carp of the same species, the mirror carp was by far the most widely accepted as a table fish – it satisfied the breeder, the grower and the housewife.

Because of the various climatic conditions in Europe there were variations in colour and in shape. Producers began competing with each other, each claiming superior fish. Fish became known by the district or county they were bred in. For example, the Galizier carp which came from Silesia in Poland, the Ukraine frame carp, the Bohemian carp from Czechoslovakia, the Franken carp from Dinkelsbuehl and the Aischgruender carp from Upper and Middle Franken in Bavaria. The Aischgruender carp is named after the River Aisch which flows into the River Main and thence into the Rhine.

Carp farming in England did not develop in the same way as on the continent. Though fish ponds appeared in medieval times, they were not stocked with carp. There are records in the Domesday Book of houses with attached fish ponds, such as that of Robert Malet from Yorkshire who owned piscenae (20 fishponds) for the holding of eels. These were undoubtedly stewponds which were used primarily for holding wild caught fish so that occasional meals of fresh fish could be eaten during the winter months. Most old monastic houses had fish ponds but the monks left no records as to how they were managed. However, these ponds must have been important since the death penalty could be imposed for the theft of fish. In fact this penalty remains on the statute books today.

The change from holding fish to fish culture was either an idea from the continent or the result of fish spawning whilst being held in the stew ponds. This was certainly the case with carp which by 1450–1500 had been introduced into England. By 1600 the techniques of breeding and rearing carp were well established and are chronicled in a book written by John Traverner called 'Certain Experiments Concerning Fish and Fruit'. This book describes in much detail the practice used for successful rearing of a number of freshwater fish, particularly carp. So good were his techniques and observations that most are still relevant for modern day carp culture, though the introduction of some modern technology has altered some of these original methods (see *Intensive carp farming in warm water* and *Induced breeding*).

Around 1800, advances in transport and marine fishing increased the availability of sea fish throughout the country. This brought about a decline in the practice of freshwater carp farming and its eventual demise. But carp had by now established themselves in our lakes and rivers. A life span of 40–50 years is not uncommon for the fish and as long as they live they continue to grow. Sports fishermen soon became aware of the merits of this large and handsome fish, particularly since the carp puts on a tremendous fight when hooked – as a result it became popular in angling circles.

What then is the difference between the table carp and the 'wildie' which is so familiar to carp fishermen? The difference can be compared to that of a pedigree dog to a mongrel, or a bacon pig to a wild boar.

Remarkably, the same species (*Cyprinus carpio* L) which started its long journey west and eventually arrived in England, simultaneously spread east through China to Japan to arrive in England much later as the koi (see chapter 16).

Over the centuries there have been differences in growing and keeping the various types of carp and these are explained in the appropriate chapters. However, breeding habits and techniques are uniform throughout all varieties.

2 Carp farming: present and future

Although not true of the UK, table carp culture accounts for about 90% of world freshwater fish culture. The remaining share comprises cultured species of salmonid, tilapia, catfish; and of lesser importance, fish such as pike, zander and tench.

The largest producer of table carp is known to be China where there are carp ponds almost everywhere; on small-holdings, in rice fields, monasteries, dams and village ponds. In short, anywhere with a surplus of suitable water. There are no figures available from China, but it is estimated that production exceeds 100,000 tonnes per year. Comparable figures are likely for Japan and the Soviet Union. In Europe – East Germany, Poland, Czechoslovakia, Hungary, Romania and Yugoslavia each produce 10,000–20,000 tonnes. West Germany, Holland, Belgium and France each produce approximately 5,000 tonnes per year.

One of the most efficient and newly developed carp farming countries of the last 20 years is Israel; producing at the present time something in the region of 30,000 tonnes annually (as well as other species such as tilapia and catfish). In recent years they have also added koi carp to their fish culture industry.

In England fish farming and in particular table carp farming had almost disappeared generally as a result of the availability of sea fish. The price of a fish and chip meal some 40 years ago was relatively inexpensive; however the situation began to change as the main fishing grounds were depleted, thus creating an opening for an alternative. Trout pellets were cheap to buy and for a time trout farming appeared to be a viable alternative. Even smoked trout began to appear on the market.

Unfortunately the price of trout pellets rapidly increased, which resulted in many small trout farmers being forced out of business as they were unable to absorb the high overheads. Another unpleasant side-effect developed for some trout farmers. Trout are carnivorous and trout pellets contain a high proportion of protein, therefore the waste water flowing back into the streams began to create pollution problems. However most of these problems have now been resolved and the trout is well established as a table fish. Furthermore intensive salmon farming is now developing rapidly in Scotland.

Up to the end of the Second World War the demand for table carp in England did not exist. It was only after the arrival of many eastern Europeans in this country, as a direct result of the war, that the traditional demand for table carp around Christmas time became apparent. This demand was met by continental food shops importing frozen carp from Poland and Czechoslovakia. All this time wild carp were present in rivers and streams. They had been living and breeding there since the Middle Ages.

There are a number of differences between the wild carp and table carp – the main one being of taste. Carp can live in poor water conditions, and many British waters are very polluted. Carp are bottom feeders and in their constant search for food they pick up all kinds of waste. In time, this feeding behaviour spoils the flesh of the fish with regard to taste and consequently they become unsuitable for the table. Carp which are reared for the table generally consume higher quality food items (supplied directly or indirectly by the farmer) which improves the quality of the flesh, resulting in a clean and pleasant taste.

As more people from eastern Europe and the Far

East came into the UK, it was apparent to the author that carp could once again be farmed for the table to fill this developing market, and soon after the first live Dinkelsbuehler table variety were imported.

At the present time about 800 tonnes of frozen or chilled table carp are imported, and about 100 tonnes are produced in Britain. If table carp was more freely available it is possible that quite a steady demand could be developed – for example it is apparent that many Chinese restaurateurs like to offer a carp dish on their menus. What then is the future for table carp farming?

In recent years great efforts have been made by Governments world-wide, in the form of grants and training schemes to stimulate production of fresh fish for their large populations. Some countries have already made very good progress in carp culture, especially where there are summer temperatures between 15 and 30°C. In Nigeria, carp can be harvested twice a year at an average weight of 0.5 kg. In the cooler parts of the world intensive carp farming is now taking place where there is a surplus of warm water from power stations and other large factory systems. It is reported that intensive carp farming is now established in parts of Siberia and northern Germany.

Newhay Fisheries began intensive carp farming on an experimental basis at a nearby power station in 1982 (see *Intensive carp farming in warm water*).

However, whether British people will take to eating carp on a large scale remains to be seen. It will surely take some time to convince the public that the taste is as good as any other table fish (in fact the taste is more fleshy than fishy.)

The disadvantages of carp eating are that it takes longer to prepare than many fish and has more bones than trout or cod. It is not easy to fillet, therefore it is a fish best suited to boiling or baking. In its favour it can be said that it is definitely much more economical to farm than any other freshwater fish. It is relatively cheap for the housewife to buy and is a very healthy fish to eat.

It is medically accepted that all the dark coloured fleshed fish such as mackerel, salmon, herring, trout and carp are much healthier from a dietary point of view than the more popular white fleshed fish such as cod and haddock. The former are high in polyunsaturates and regular consumption of such fish will result in a lowering of cholesterol in the blood. In some parts of the world where the people regularly consume these oily types of fish previously mentioned, heart diseases are almost non-existent, (for example, amongst the Eskimos, Chinese and Japanese). So with England becoming more and more health food conscious the table carp may have an important role to play; the fish is starting to appear on fishmongers' slabs. And at a time of excess agricultural production of butter, meat and wheat there is surely an argument for development of some alternative products. That is not to say that all farmers should turn to carp production, but many estates and farms have waterlogged or partly flooded waste land which although unsuitable for most purposes, could be converted into carp ponds.

This expansion would echo developments in other parts of Europe. With the increasing number of overseas visitors and immigrants from countries where carp is part of the customary diet there should, in my opinion, be a greater interest in carp farming.

3 General information

What is a table carp?

The definition of a table carp is: a pedigree carp which is produced under controlled conditions. These fish are carefully selected; the breeder tries to ensure that the fish are fast growing, show a good flesh to bone ratio, that the head is as small as possible, that the flesh has a clear taste and does not contain too much fat, and that the fish is sexually immature. Generally, table carp are 3 summers or 2½ years old. At this age the flesh is firm and there is little to discard as inedible, such as roe and milt.

By selective breeding over a period of time all offspring become identical producing a pedigree carp in the same way as has occurred with chickens, pigs, cattle, horses *etc.*

As mentioned earlier, over the centuries certain characteristics such as body length, colour, and scale formation of the carp have been altered by selective breeding. On the European continent this has resulted in varieties like the Dinkelsbuehler carp which was the first to be imported to England. It is well suited to the British climate and has been successfully bred as a table fish for the past 15 years. From these original stocks the Newhay table carp has been developed.

Information for the beginner

Generally speaking the carp can tolerate almost any conditions where other freshwater fish survive. But the economics of farming carp for the table dictate that faster growth rates must be achieved than those which would normally occur in the wild. Carp are basically warm water fish. In wintertime they hibernate, therefore no feeding occurs and very little attention is needed. A good feeding temperature is 15°C (59°F), and with it, rapid growth commences. C1 (carp of one season) start feeding at 6°C (42°F). C2 and C3 begin at 10°C (50°F). But at these temperatures food is poorly digested, particularly when the diet consists mainly of water plants and algae.

In choosing a site for carp ponds construction the following are important criteria:

- availability of an abundant supply of suitable water;
- a sheltered position (natural valley);
- trees – preferably coniferous (for shelter) adjacent to site;
- sufficient space for pond development on one or both sides of water supply (see *Fig 4*).

Starting from scratch It is tempting to think that one can simply dig a large hole in the ground, fill it with water to grow carp and

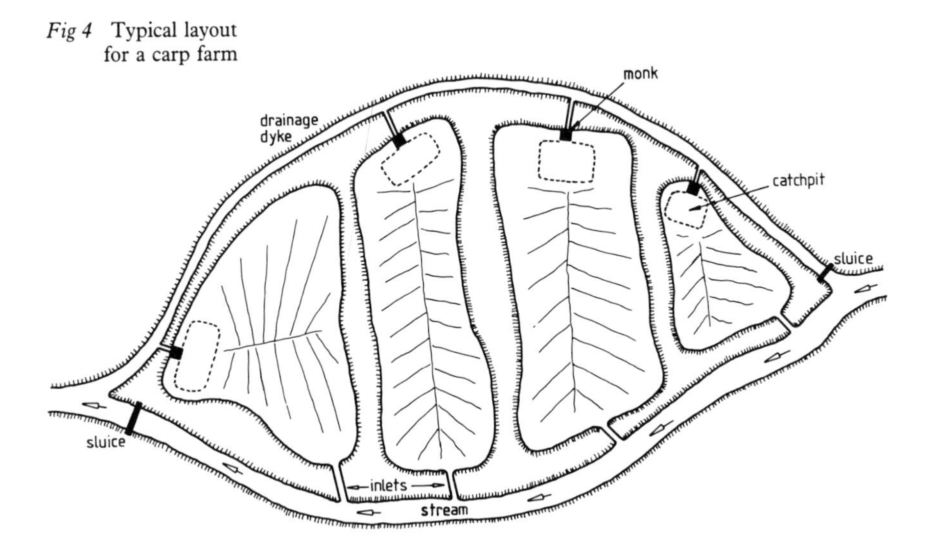

Fig 4 Typical layout for a carp farm

quickly make a handsome profit, but of course this is not the case.

The first step to take is to approach the local water authority to obtain permission to extract water from an appropriate source. A stocking permit is also required before introducing fish to any freshwater areas in Britain.

To make a farm viable the ponds should not be smaller in size than 0.14 ha (⅓ acre). There must be sufficient water available to fill up the pond and to keep it topped up during dry periods. It is most unwise to begin digging before the water supply has been tested for quality and suitability for carp farming. Again the water authorities are most helpful in this respect and generally will test the water free of charge.

The next step is to obtain planning permission from the local council, which may be a difficult and lengthy process. Once granted, estimated costs for the digging of ponds can be obtained from contractors. These should include the cost of installing monks and catch-pits. It should be remembered that there will be large amounts of surplus soil, some of which can be used for the building of banks. It may also be possible to dispose of some around the building areas for uses such as levelling. The more soil disposal can be dealt with on or near the site, the lower will be the over all cost. Loading and carting away excess soil can increase dramatically and even double the cost of excavation. However, there are known cases where the surplus soil has been sold and the whole developing cost has been covered in this way.

It is advisable to check on the availability of manure and inexpensive sources of food for the carp, such as waste from farms, bakeries, hotels and flour mills, (see section on building a carp pond).

Adapting existing ponds

In any pond where the water quality is reasonably good fish are likely to be present having been introduced by man or carried there by animals, particularly birds.

Many ponds are already in use by clubs and

syndicates for angling purposes. Many of them are deep enough to be stocked with trout (particularly if spring fed) providing summer temperatures do not exceed 15–18°C (54–64°F). But there are still a large number of ponds which could be used for carp farming; most suitable are those which can be drained or pumped empty. Existing ponds may be equally as good as those which are purpose built, particularly if they are in a sheltered position with little water movement in or out and where a net can be easily operated. Problems arise where there are predatory fish present in the pond (*eg* eel, pike, zander and perch) as it is difficult to ensure that such fish will be removed by netting. One solution is for fry and C1 fish to be grown on in smaller separate ponds and then transferred at C2 size to the original pond; the carp by that time are too large to be at risk from attack by the predatory fish.

The disadvantage in operating ponds which cannot be either drained or pumped empty is that harvesting has to be accomplished by more labour intensive methods such as netting or electric fishing, and there is no guarantee that all the fish will be caught. If after harvesting, some of the fish remain in the pond these larger specimens will compete for food with the next season's introduction of smaller fish. Consequently the smaller ones would be deprived of the amount of food they require due to the proportionate intake of food by the larger fish. Also it should be noted that as each pond can only support a certain number of fish, overstocking may result in a serious decline in the zooplankton population – in which case additional food must be supplied to the fish.

Wherever this form of table carp farming takes place there is always a quantity of large fish in the catch which are not acceptable as table fish. The average weight of one C3 table carp is about 0.9 kg (2 lb); any larger fish can be sold very lucratively to the angling market.

The age of a carp can easily be estimated by studying

the growth rings on the scales – similar to the way in which one determines the age of a tree by counting the rings in the trunk (see *Fig* 5).

Building a carp pond

A carp pond must fulfil several roles. It represents the fish's whole world, including its living quarters, kitchen and toilet; and from the farmer's viewpoint must supply the carp with suitable conditions for maximum growth. Ponds should be built to last; to be easily drained and harvested rapidly without damage to the fish. In principle, on any site where there is sufficient water available, a carp pond can be excavated. It is known that better growth rates occur in ponds dug in fertile land but it is likely that carp farming may still be feasible in areas where soils are poor but where there is a plentiful supply of good quality water. However, a prospective site must be thoroughly surveyed to ensure that sufficient water is

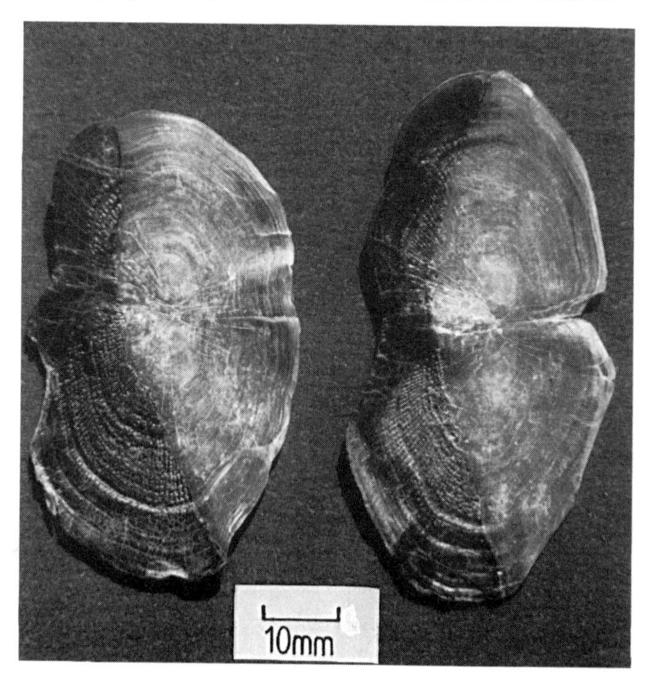

Fig 5 Scales from a 10-year-old mirror carp

available, that the pond can be drained and that a degree of shelter exists: (better growth rates are achieved in warmer, undisturbed waters – see section on starting from scratch).

It is necessary to have levels taken by an optical levelling instrument such as a theodolite. The deepest end of the pond (monk end) should have a depth of water between 120 and 140 cm (4½ ft). The pond should slope upwards to the shallow end at a rate of 2 to 3 m per 1000 m length of pond. The pond slopes from both sides towards the centre at the same rate leaving a groove in the middle about 20–25 cm (8–10 in) wide (*Fig* 6).

It is preferable to have several smaller ponds rather than one large one, but they should not be less than 0.14 ha (⅓ acre) (except fry or Dubish ponds). To construct deeper ponds than those mentioned is not advisable as the sun does not penetrate into deeper water, hence the water remains relatively cool and the growth of plankton is adversely affected. The bank should be about 50–100 cm (approx. 2–3 ft) above the water level so that the whole depth from the bottom of the pond to the top of the bank is approximately 2 m

Fig 6 Topsoil returned to pond

(6½ ft). The width of the top of the bank should be about 3 m (10 ft) – with larger ponds, where the soil is somewhat porous, 3–6 m (10–20 ft). Soil on the banks should be compacted, this is best done with the shovel of a digger. If the soil is very dry it is advisable to wet it before doing so.

Stabilising the soil and the maturing of the bank means that sometime later there will be some settling, this can be as much as one fifth of the bank depth and will have to be allowed for when making calculations. Before any digging begins, it should be remembered that about 25 cm (10 in) of topsoil will have to be removed and placed in a separate heap; on completion of the pond this soil is returned to the pond to supply nutrients required for the growth of plankton. In *Fig 7* a cross-section of a pond bank is shown; in areas where the soil is somewhat porous a clay core may be incorporated into pond banks.

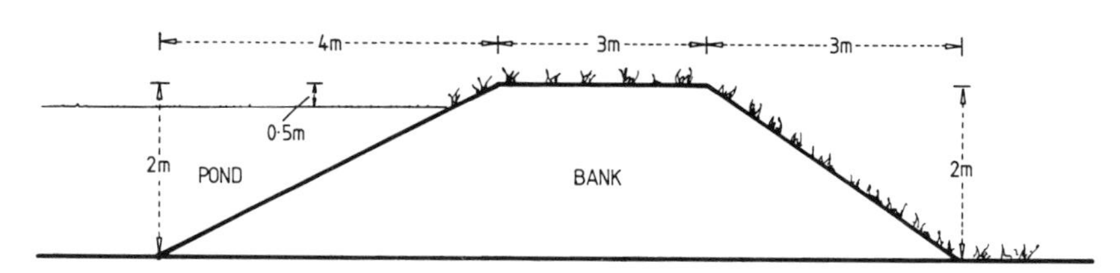

Fig 7 Section through pond bank

Fig 8 Waterlogged grassland before commencement of pond building

Fig 9 Pond building in progress

Fig 10 Finished pond

The wintering pond In the winter months when water temperatures may fall below 10°C (50°F) carp from C2 size upwards gradually stop feeding and prepare to over-winter (C1 carp continue to feed until the temperature falls to 6°C /43°F). The carp makes a small dent on the bottom of the pond and with head down at an angle of about 45 degrees begins to hibernate. The carp cluster together in large groups for this dormant period.

If in exposed situations the wintering ponds should have a minimum depth of 140–200 cm (4–6½ ft). But such ponds should preferably be in a quiet position away from all forms of vibration such as from roads or large factories. Ideally there should be a small amount of water flowing through the ponds at a rate of 680–910 *l* (150–200 gal) per hour. Too much flow through is not advisable as this will disturb the fish, but if this is not possible the pond should be stocked with approximately half the quantity of fish (see *Stocking the wintering pond*).

The monk

A good carp pond can not function to its maximum capacity without a monk. This is basically a sluice named after the monks who invented it and fulfils a number of roles:

- sealing the pond completely thereby ensuring that no water is allowed to escape;
- controlling the depth of the pond water;
- together with the catch-pit – ensures quicker and more efficient harvesting;
- keeps the pond empty throughout the wintering period.

In the past monks were fashioned mainly from timber and, as they had to work correctly, were made by craftsmen, and consequently were very expensive. Later monks were made from reinforced concrete; with the advantage that they could be cast from a mould. The disadvantage was that when used on unstable ground the weight of the monk (approximately 1 tonne) could cause it to move possibly resulting in a cracked outlet pipe and hence a leak. Also a prolonged spell of frost may crack a concrete monk (*Fig 11*).

There is now a monk available which is constructed from fibreglass. It has many advantages; it is competitively priced; will not crack, is easily installed and can be fairly easily removed and, if necessary,

Fig 11 Concrete monk

transferred to another pond even after many years of use (see *Fig 12*).

A monk has three sets of grooves to take screens or boards which control the pond water level. The groove nearest to the pond side is used to hold a screen which slides into the bottom of the pond. The size of the mesh used in the screen is determined by the size of the fish to be harvested. Above this screen a board is inserted which goes up to the top of the monk. This

Fig 12 Fibreglass monk installed in pond bank

means that any water draining through the monk will have to go through the screen at the bottom thus preventing the fish from escaping. Boards are inserted into the two remaining grooves. The gap between these boards is best filled with wet sawdust which will swell and make a complete seal. In the absence of sawdust, clay or even mud can be used. The top level of these boards controls the water level of the pond. A drainage pipe which is attached to the rear of the monk, passes through the bank into a drainage channel, dyke or stream. To drain the ponds slowly a pair of boards are removed and the water begins to flow out until the new level is reached. A further pair of boards are then removed and this procedure is repeated until the pond is empty.

Catch-pit inside the pond

Catch or harvesting pits are usually built into the pond during construction. These are simple pits dug into the bottom of the pond below the level of the monk. The pit remains filled with water when the pond is drained. As the pond empties the carp swim with the flow and eventually find their way to the catch-pit. Then they can be easily removed, or alternatively held for a short time, provided there is a water flow through the catch-pit. The pit can be simply a deepening in the bottom of the pond or can be purpose built from concrete blocks to provide a permanent structure. A good and less expensive pit uses railway sleepers, these last for many years. The size of such a catch-pit for a pond 0.2 ha (½ acre) size would be about 6 × 3 m (20 × 10 ft) and 0.6 m (2 ft) deep (*Fig 13*).

Catch-pit outside the pond

There is another form of catch-pit which is built outside the pond on the outer side of the bank and is more expensive to construct. *Figure 14* demonstrates the workings of this type of catch-pit.

When the pond is to be harvested, the screen on the large shutter is removed completely. The shutters in the remaining two grooves are taken out as explained in the other forms of harvesting. The carp follows the water flow and as there is no screen, will pass through the monk, through the connector pipe and into the catch-pit. The water will pass through the catch-pit but the fish are prevented from following because of the screen between the catch-pit and the dyke.

When the pond is empty and all fish are in the catch-pit, the water supply from stream or dyke is

Fig 13 Catch-pit inside pond

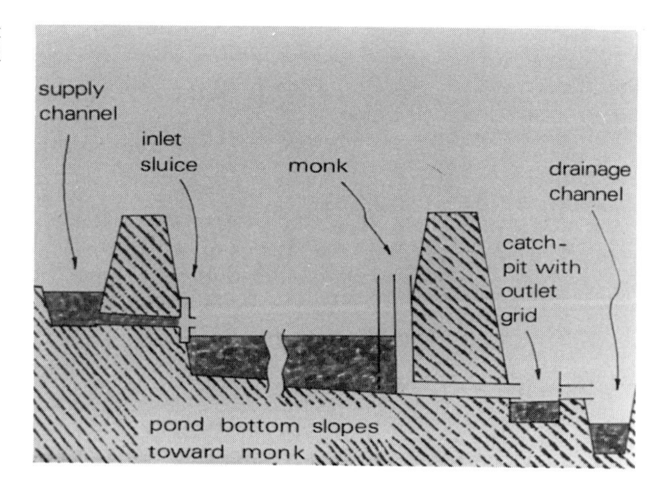

Fig 14 Catch-pit outside pond

directed through the pit providing the fish with sufficient oxygen. In winter at low temperatures all the fish from the pond can be kept alive in the catch-pit as the through flow of water maintains a supply of oxygen.

Grooves can be built in to hold screens of various sizes and fish are graded in this way (*Fig 15*). Lockable lids can be fitted to the catch-pits to secure the fish from unauthorised removal. Most large carp farms harvest their fish in this way. In more mechanised operations conveyor belts, especially made for fish farms, are set into the catch-pits and connected directly to the waiting transport (*Fig 16*).

Water supply

The water supply to the pond is of vital importance. Not only must the water quality be reasonably good for carp farming, but the supply must be maintained throughout the year. In the past many would-be carp farmers have greatly underestimated the quantity of water required. The following conversation has often been repeated: Question 'Has your stream or dyke ever dried out in a prolonged drought?' Answer 'Well, that happens perhaps only every ten years or so, and when it happens I can always use my garden hose to top up the pond'. If the pond in question is of a size of approximately 0.4 ha (1 acre) then it is often surprising to the beginner how much pond water is lost in the summer months by seepage and evaporation. In such cases it is quite impossible to fill the pond by means of a hose pipe.

Ideally the water supply should come from a stream which has a gradient of about 3 pro mill, *ie* within a horizontal distance of 1,000 metres the water drops a vertical distance of three metres, thus if the pond is 300 metres long with a monk at its lower end, the pond could be filled from the upper end to a depth of 0.9 metres. This would not be quite deep enough in this instance as the aim would be for a depth of 1–1.40 m.

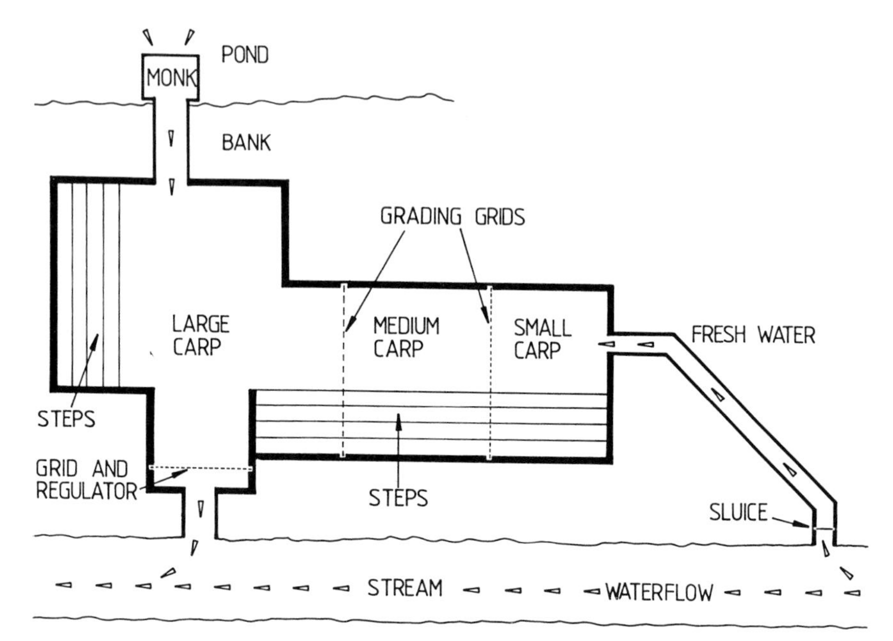

Fig 15 Catch-pit with grading grids

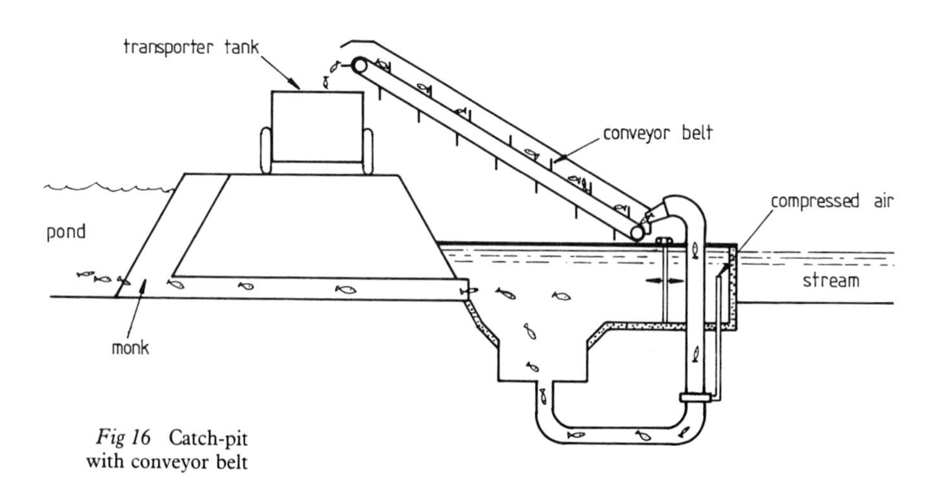

Fig 16 Catch-pit with conveyor belt

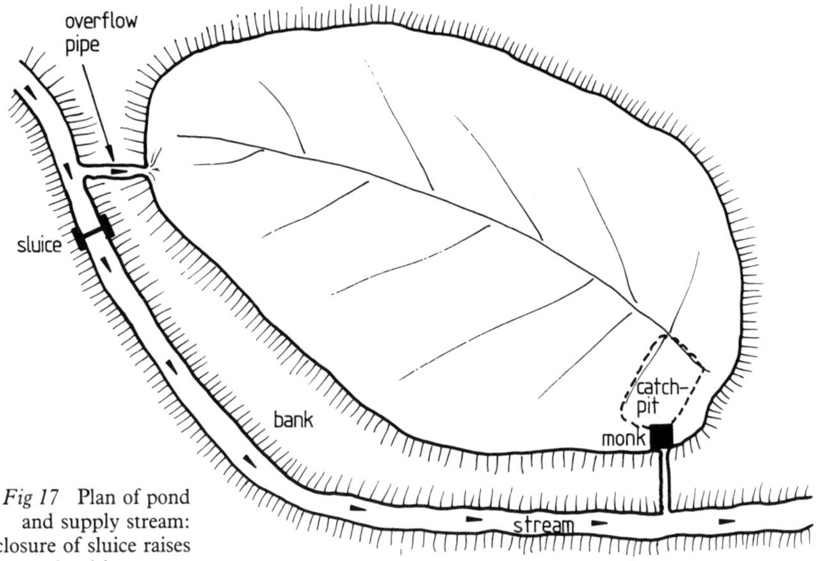

Fig 17 Plan of pond and supply stream: closure of sluice raises water level in stream, overflow water fills pond

It may be possible to raise the water level from the upper end by means of a simple sluice (guillotine sluice) during the time that the pond needs to be filled (*Fig 17*).

It could be argued that for the pond level to fall a given number of centimetres is not a serious matter, but this is incorrect – as in most other industries, the practices involved in carp farming are tried and tested, and if short cuts are taken, the consequences are often disastrous. If the pond is stocked to its maximum limit the water level must be kept at a predetermined level or the required biological balance within the pond can not be maintained. Loss of large quantities of water in the summer months, if not replaced, results in higher water temperatures, the carp become more active and hence use more oxygen and food to replace lost energy; an increase in waste products follows which further depletes oxygen levels. This sequence of events may lead to a situation known as 'carp uproar' in which fish swim to the surface in an attempt to gulp atmospheric air, if these conditions persist, mass deaths amongst the fish will occur. What then must the carp farmer know in respect of a water supply? With, say; a 0.4 ha (1 acre) pond, the evaporation and seepage in the summer months accounts for something in the region of 40–60 cm (16–24 in) from the pond surface. In one dry summer month alone evaporation can account for 20 cm (8 in). This means that taking seepage into account also; approximately 820,000 *l* (180,000 gal) of water can be lost. To replace this quantity for this one month alone would need an inflow of water at the rate of 23 litres (40 pints) per minute, day and night for a whole month. To avoid such problems the water supply must be very carefully investigated before carp farming can begin.

Should it be found that water supply is less than adequate or the water level in the supply stream can not be raised it may be possible to collect more water by widening and/or deepening the supply stream; water can then be pumped into the pond by mechanical means. A stewpond alongside the stream to collect and store water in more plentiful times, or possibly a nearby borehole or lake could be used to supplement existing supplies.

In many low lying areas, the water table is sufficiently high that the water supply is almost constant both in winter and summer. It may be possible to dig ponds of reasonable size in such areas. These ponds would be harvested by pumping the water out; pond liming must then take place immediately, before the pond refills itself naturally. It is also possible, and even desirable, to build ponds in areas alongside rivers which flood during prolonged rainfall or sudden downpours – such areas are of little value to farmers, except perhaps as poor quality grazing land. If carp ponds are constructed in such a situation the banks should be built high enough to prevent flood water escaping over the top of the bank. On the river side of the ponds a large grid is built into the pond bank – this determines the depth of the pond water. In the case of flooding, water flows through the grid raising the level of water in the pond, and will flow out again as the flood recedes. The grid will prevent carp from escaping and will also prevent unwanted fish and debris, carried on the flow, from entering the pond. In cases where the grid is the only source of water inflow it is advisable to have a deep stewpond adjacent to the ponds for topping up purposes. Ponds like those described are operating successfully in Britain. It must be mentioned however that water authorities will not readily grant permission for pond construction in such areas, and the would-be carp farmer must prove that the flood area would remain constant. Otherwise there would be a valid argument that by interfering with one flood area a flood elsewhere would be created.

Summary The whole process of building a good, economically functional, carp pond may appear somewhat complicated to the beginner. Problems do arise but if a landowner or farmer is considering developing two or three medium sized ponds, the information in this book should be sufficient, but it is important not to cut

Fig 18 Grid for floodwater

corners. If the digging has to be contracted out at least three estimates should be obtained. In one development three ponds were to be constructed and four contractors were invited to quote. The quotations varied enormously, the highest being four times more than the lowest. In cases where the area to be developed is 4 ha (10 acres) or more, it is advisable to seek the service of a consultant to ensure that nothing is overlooked.

The pond water

It is generally believed that clean pond water is pure but quite categorically it must be said that there is no such thing as pure pond water. Even rain water is far from pure; before it enters a pond it collects a variety of chemicals, and as rain water constitutes quite a large percentage of the water in ponds, it follows that chemicals will accumulate there. A particularly dangerous situation for pond farmers occurs after a period of drought. When the welcome downpour takes place the water entering ponds may contain toxic materials which have accumulated during the dry period. This is particularly the case where ponds are sited close to power stations or industrial installations. Environmental damage caused by acid rain from industrial pollution is becoming increasingly common. Similar situations may occur in winter time when the ponds are frozen and there is also an accumulation of snow or drifted snow on the ponds. Should there be a sudden thaw, the water may become toxic due to the impurities which have collected in the snow and ice; in addition, melted snow contains little or no oxygen, and this is likely to have an adverse effect on pond life. Fortunately, such situations do not often occur, but if they do, the carp farmer must be prepared to aerate the ponds by mechanical means. In case of snow drifts on the ponds. it is advisable to remove as much of this as possible before the thaw.

It is not only the carp which needs oxygen but the whole pond requires it in order to function as a biological unit. As water flows down a stream or river, oxygen penetrates and saturates the water. In a carp pond the oxygenation procedure is much slower and occurs by the action of wind at the surface, by the action of fish swimming and even by the movement of insects. Additionally, the water is oxygen enriched by water plants, including algae and plankton. Water plants give out oxygen during the day, but take a certain amount of this in again during the night. The following table indicates how the oxygen content of the water varies in a typical pond both by day and by night:

Time of day	*Dissolved oxygen mg/l*
2 am	7.8
6 am	5.3
10 am	5.7
2 pm	7.4
6 pm	9.1
10 pm	10.2

Just prior to dawn is the time when oxygen levels in the pond are at their lowest. The oxygen content of the water is important because fish, like land animals, require an adequate supply of oxygen to breathe but in the case of fish, their oxygen is dissolved in the water. The level of dissolved oxygen should therefore be as high as possible to ensure that the fish receive an adequate supply.

Water can only hold a limited amount of oxygen under normal conditions – the saturation level. This level is influenced by water temperature and the table on *page 48* shows dissolved oxygen saturation in relation to water temperature.

The table shows that as water temperature rises, the amount of oxygen that can be carried in solution decreases. Unfortunately, as this occurs the fish become more active and require more oxygen. The fish farmer should be aware that at times of high water temperature, problems of reduced oxygen levels can occur.

Temperature °C	*Dissolved oxygen mg/l*
2	13.40
4	12.70
6	12.06
8	11.47
10	10.92
12	10.43
14	9.98
16	9.56
18	9.18
20	8.84
22	8.53
24	8.25
26	7.99
28	7.75
30	7.53

Oxygen levels are measured in either milligrams per litre (mg/*l*) or parts per million (ppm). One milligram of oxygen per litre means that one milligram of oxygen is dissolved in one litre of water; and is approximately equivalent to one part per million.

Every species of fish has its own particular requirements for minimum oxygen content of the water. For carp, the oxygen level should not fall below 3 mg/*l* during the production season; though this minimum level will only ensure survival, not growth. During the winter months, when the fish are in a state of hibernation, this minimum level reduces to 2–2.5 mg/*l*. For comparison, trout have a minimum requirement of 10 mg/*l*. A reduction in the oxygen content below these minimum levels will quickly result in fish mortalities.

During the summer, when temperatures are at their peak, it is best to try to keep the oxygen levels of carp ponds as high as possible. The minimum desirable oxygen level for growth is about 6 mg/*l*. If values below this are measured then some corrective action should be taken, such as mechanical aeration or addition of fresh water.

There exists, if correctly balanced, a steady exchange programme between fish and plants, which works as follows: by day the fish breathes in oxygen and breathes out carbon dioxide, which is received by plants and converted to oxygen. During the hours of darkness both fish and plants take in oxygen and produce carbon dioxide but as fish are less active at night, on balance the process is beneficial to the fish. It has to be remembered that in summer time, water temperatures can be very high and the oxygen demand of both fish and plants is also high, therefore the most critical time for oxygen starvation is in the hours around dawn.

pH Another important factor that dictates the productivity of the pond is the pH value. pH is a measure of the acidity or alkalinity of the water and is measured on a scale of 1 to 14. Neutral water has a pH of 7, values below 7 indicate acid conditions and above 7 alkaline. The pH of natural waters lies between 4 and 9 and depends on the source of the water and the nature of the surrounding soil and rocks over which it passes.

Water which is to be used for fish culture should have a pH value of between 6.7 and 8.5. The ideal level for carp is about 7.5–8.2 as these values promote the best conditions for the growth of food organisms. The following table shows the effects of a range of pH values on fish growth.

pH values	
4	– Death
5, 6	– Slow growth and no reproduction
7, 8	– Good growth
9, 10	– Slow growth and no reproduction
11	– Death

The pH of natural waters can change very quickly, especially during times of heavy rain when acids from the soil are washed into water sources. Low pH values can also be experienced from the rain itself when it is contaminated by industrial pollutants which produce what is now known as acid rain.

Hardness or acid binding capacity

Hardness is a measure of the total soluble salts in the water. These are usually calcium and magnesium compounds and they are beneficial to fish and to the plankton which the fish utilise for food.

Varying degrees of hardness have been assigned to different water types: 0.75 mg/*l* – soft water, 0.75–150 mg/*l* – moderately hard, 150–300 mg/*l* – hard, and over 300 mg/*l* – very hard. It is thought that as long as the hardness value of pond water is over 20 mg/*l*, then no deleterious effects will be seen. Hence it is evident that in most water types hardness levels will be adequate for fish.

Ammonia

Although not important when considering whether a water supply is suitable for fish culture, monitoring of ammonia levels becomes a priority once a pond is stocked with fish, especially when supplementary feeds are in use.

Ammonia is formed in the pond water from fish excrement, waste food and organic fertilisers as well as by microbial decay of other nitrogenous compounds. In ponds where fish densities are kept too high, ammonia concentrations can quickly reach undesirable levels. The chemical can occur in two forms in water, as free ammonia in solution or as the ammonium ion. The principal factor that influences which of the two forms occur is the pH of the water. Above pH 8.0 the free ammonia form is favoured, and below pH 7.0 the ammonium ion. Free ammonia is toxic to fish at quite low levels. Small fish can not tolerate levels of above 0.2 mg/*l*, large carp can withstand levels of 1 mg/*l*, but only for short periods.

The safe limit for all fish is less than 0.1 mg/*l*. The ammonium ion is harmless to fish.

If ammonia is detected in the pond water then action must be taken immediately; particularly in alkaline water, all supplementary feeding should be stopped. Attempts should be made to trace the source of the chemical, the usual cause being overfeeding or excess fertilisation.

Associated with ammonia in the pond are nitrites and nitrates. These are formed as a result of the microbial breakdown of ammonia by the natural processes within the pond. The breakdown occurs from ammonia to nitrite and then nitrate. Nitrite is toxic to fish, levels of 0.1–0.25 mg/*l* are dangerous and 0.5 mg/*l* can be lethal. Fortunately nitrite is usually oxidised very quickly to nitrate.

Nitrate is not toxic unless present in very high concentrations (in which case the water would probably be so polluted that the fish would be already dead); at normal levels it is an important nutrient source.

Testing water quality

The chemical composition of pond water can be determined using specialised testing equipment. Both chemical and electronic apparatus is available, the latter is expensive, but very accurate. Chemical testing is somewhat less precise, but adequate for carp farming purposes. Equipment can be bought as a complete set or as separate testing kits for oxygen, pH, ammonia etc. For a very rough assessment of water quality there are certain observations which can be noted. For example, shortage of oxygen will result in fish appearing at the surface attempting to gulp down atmospheric air; water snails may also be visible near the surface. Mosquito larvae, *Daphnia* and cyclops can all be found in normal pond conditions from spring until early autumn – if these species are absent at these times, it is a further indication of water quality problems. Some larvae however, have a preference for poor water conditions, for instance the larvae of the

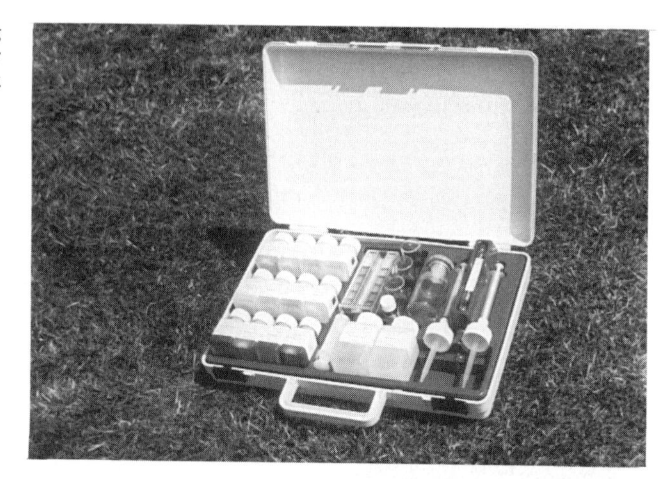

Fig 19 Water testing kit manufactured by Merck

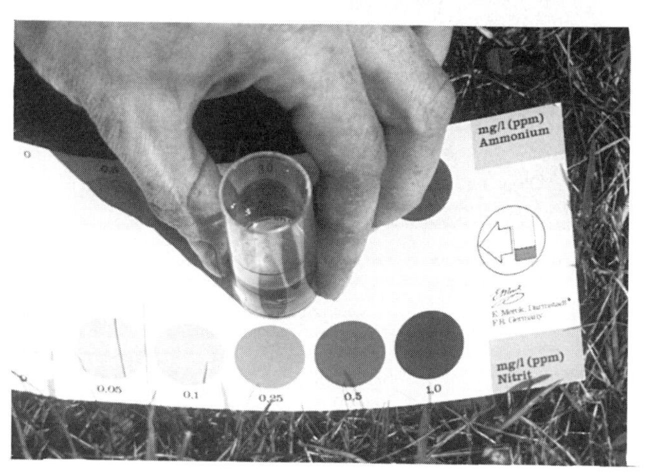

Fig 20 Testing for nitrite

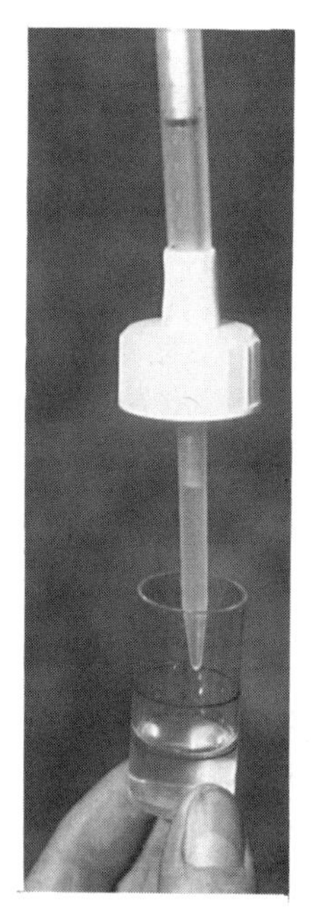

Fig 21 Testing for oxygen

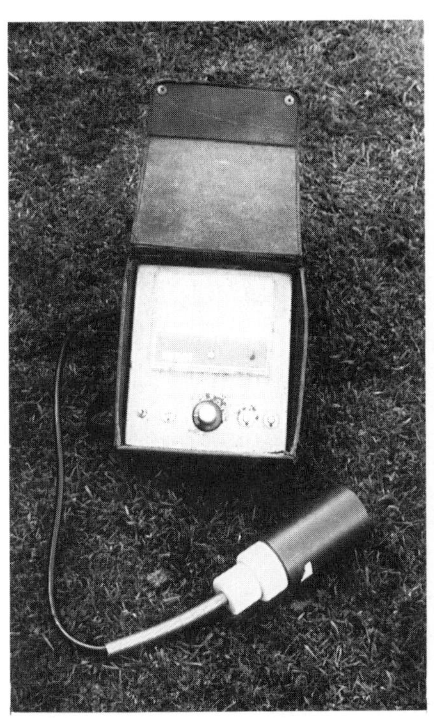

Fig 22 Oxygen meters

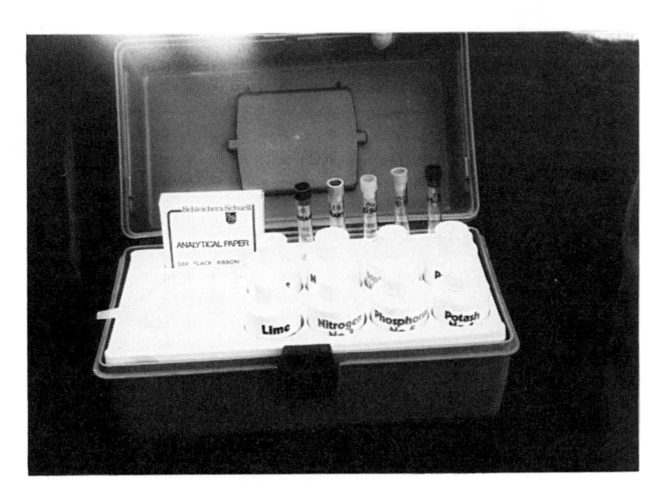

Fig 23 Test set for soil analysis

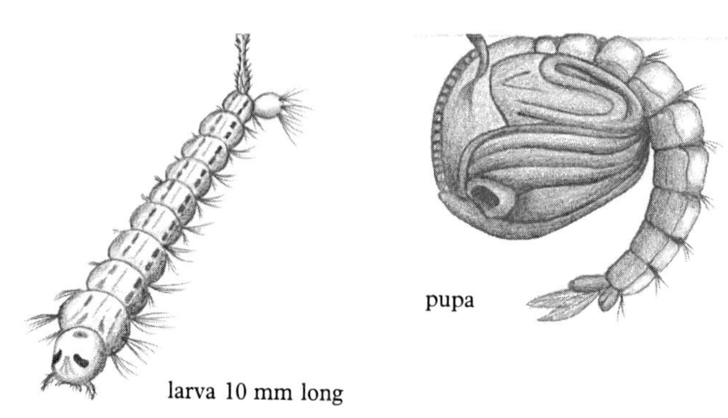

Fig 24 Mosquito (*Culex pipiens*)

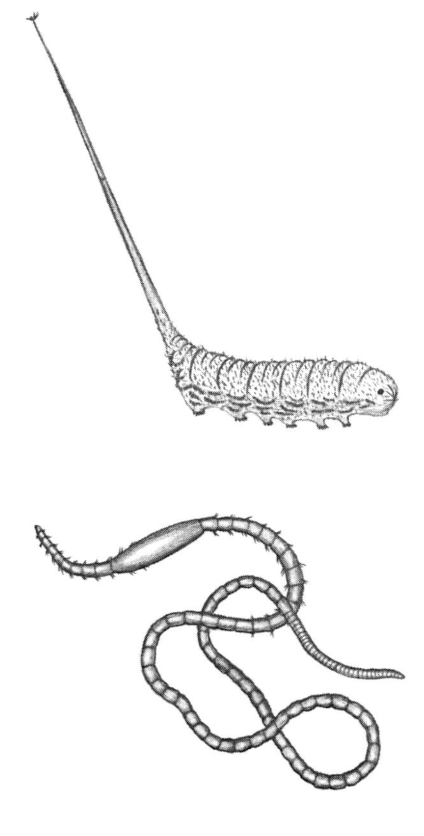

Fig 25 Rat-tailed maggot – hoverfly larva (*Eristalis* sp) up to 20 mm long

Fig 26 Sludge worm (*Tubifex tubifex*) up to 90 mm long

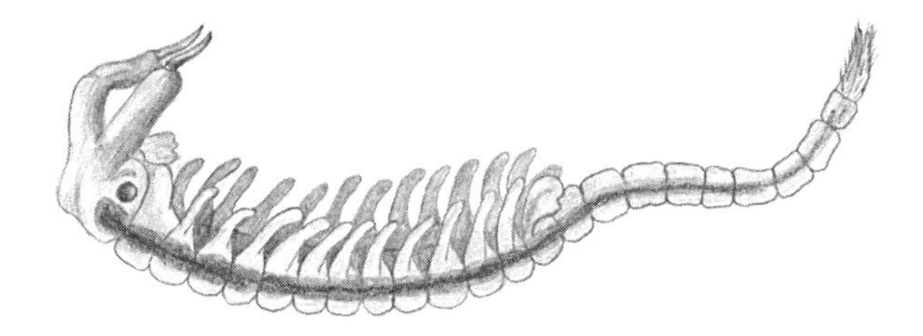

Fig 27 Fairy shrimp (*Chirocephalus* sp) up to 10 mm long

hover fly (*Eristalis* sp). These rat-tailed maggots inhabit polluted water and can breathe atmospheric air through a telescopic tube (see *Fig 25*). Where these larvae are present, it can be assumed that the water conditions are poor. The same applies to water bodies where large numbers of sludge worms (*Tubifex*) (*Fig 26*) or fairy shrimps (*Chirocephalus* spp) occur (*Fig 27*).

Water plants may also indicate certain conditions. For instance, water-cress in a pond tends to suggest good water quality as the plant likes plenty of oxygen and grows best in moving water. Canadian pondweed and duckweed favour alkaline water; whereas the presence of cotton grass, bog cotton and water horsetail suggest acid conditions.

A careful inspection of underwater plants should reveal foliage which looks fresh with no traces of fungus on stems or leaves but growth should not be so excessive that when stirred with a stick, any noticeably unpleasant smell is emitted.

Looking through a sample of pond water will give some indication of the water quality, but as previously mentioned, it can not always be assumed that clear water is necessarily uncontaminated. Water which has a greeenish tinge is generally acceptable. The green colour comes from the presence of plankton which is the basis of the food chain and only forms where favourable conditions exist.

Too much iron in the water will produce dirty red looking plants or soil and if present in excessive amounts, an oily film forms on the water. When breaking the film with a stick the colours of the rainbow can be seen on the pond surface.

Green algae, in moderation, in the pond is a good sign. But blue colouration signifies the presence of blue-green algae – suggesting unfavourable water conditions.

These indications are a rough guide only to water conditions. It is far safer, and essential for the serious carp farmer, to have proper water testing equipment.

Pollution Heavy fish losses occur every year as a result of chemical pollution entering water stocked with fish, either accidentally or as a result of carelessness. The problem is more serious in the carp pond than in rivers or streams, because the water flow in rivers and streams has a diluting effect and therefore the danger may pass before any serious damage occurs. That is not so in a carp pond where there is little or no movement of the water.

Arable farmers are using more and more inorganic fertilisers, these accumulate in water bodies and are capable of killing fish if levels become too high. It is not only fertiliser residues which are dangerous for carp; pesticides and herbicides are usually toxic and even if not, many form a film on the surface of the pond. This creates a barrier between pond water and air and as a result the oxygen content of the water then falls rapidly and the fish have difficulty breathing. They will be seen to come to the surface and try to gulp down atmospheric air; by doing so, a certain amount of the chemical film passes into the gills causing inflammation, which leads to death.

Aerial spraying represents the greatest potential threat to the carp farmer's ponds, and it is worthwhile to ascertain the identity of firms who are likely to be spraying in the pond area. Liaising with the spraying companies will ensure that spraying takes place only when the wind is in a favourable direction, to safeguard the pond from its lethal effects.

After chemical spraying has taken place, a sudden downpour of rain, for instance, can result in rapid contamination of a pond. Surface run-off from the newly sprayed land enters the pond, taking with it the potentially toxic substance which can quickly have an adverse effect on the fish population, as previously described. This situation can be avoided by the construction of dykes which divert excess water away from the pond.

Topping up the ponds after a rainy period should only take place after the water has been tested, as land drains can also carry toxic matter into the supply stream or dyke.

4 Plants in the carp pond

Even in a newly dug pond signs of plant life soon appear and the carp farmer must learn how to deal with them. The following may cause problems:

- suspended microscopic plants, plankton and phytoplankton;
- 'soft' water plants which grow submerged or with leaves floating on the surface of the water;
- 'hard' water plants, where the roots establish themselves in the water and the foliage grows above the water surface.

Plankton

When observing a drop of water under a microscope, a surprising world of life is revealed. Many of the organisms can be seen to be moving freely about. These tiny creatures are part of the animal kingdom, and include (micro-organisms such as rotifers, protozoa *etc*), but most noticeable are the tiny stick-like objects – phytoplankton, which because they contain chlorophyll are green in colour. Chlorophyll is a pigment which, by using energy in the form of sunlight, can convert carbon dioxide and water into carbohydrate, with oxygen given off as a by-product – a process known as photosynthesis. In turn, carbohydrate provides a source of available energy for growth, reproduction *etc*. As sunlight penetrates a pond the phytoplankton grows so rapidly that small patches appear, growing in the shallow end of the pond and on some of the larger plants. This phytoplankton supplies oxygen to the pond water which encourages the growth of zooplankton, and provides excellent food for the carp.

As more sunlight penetrates the pond algae rise to the surface forming large floating islands which, particularly if the water has a high pH value, will soon cover the surface of the pond. The result of this occurrence is a drastic reduction in oxygen and quite soon the entire pond fauna, including the fish, are in danger. Long before this happens the carp farmer must take action, and the best time to do this is early on. When the large floating patches of algae are blown by the wind to the side of the pond they are easily removed by raking out. An alternative method employs a plank with a rope at each end; by pulling the plank, which preferably has some nails half driven into the underside, the algae is dragged to the edge of the pond where it can be removed. Algal growth can also be inhibited by active disturbance of the bottom sediment (*eg* by wading) thus clouding the water and reducing the depth to which sunlight can penetrate. This control mechanism can occur naturally in some ponds but is dependant on water pH and the size of fish stocked. Large carp dig for food in the bottom of the pond and this action clouds the water. Smaller fish are not strong enough to do so.

A chemical is available to treat algae problems, this is copper sulphate. The quantity to use is 11.1 kg per hectare (4½ kg to one acre) of pond surface. The copper sulphate should be dissolved in water and then sprayed over the water surface. This method should only be used when other methods have failed, and with great care as it is toxic to fish, invertebrates and other animals.

'Soft' water plants

These plants are acceptable in moderation in a pond as they oxygenate the water, and their soft leaves, seeds

Fig 28 Canadian pondweed (*Elodea canadensis*) 5–15 cm long

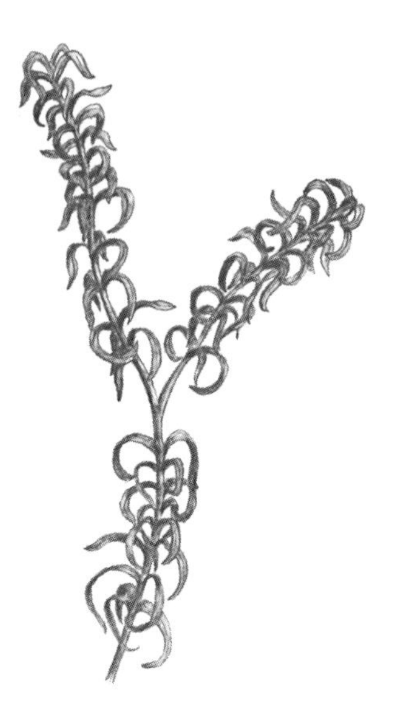

and roots provide a source of food for the carp. In certain cases, however, the growing conditions are so favourable for some species, that if not controlled, the biological balance of the pond is disrupted and fish can be lost in a very short time. Some of these unsuitable plants can have an adverse effect upon carp farming. One such plant is *Elodea canadensis* commonly known as Canadian pondweed (*Fig 28*). This plant is easily recognised, having three leaves round the stem. As the latin name indicates it is an import from Canada. The strange thing is that in some situations it can disappear as quickly as it grows, if certain minerals, essential to is growth, become exhausted. The plant can withstand extremes of temperatures, but it is susceptible to a hard frost on a drained pond. Therefore wintering a pond should get rid of this unwanted plant. C2 and C3 carp will feed on the plant initially, but if it becomes too thick, the carp will avoid it completely. Once again, clouding of the pond water will restrict its growth, and this treatment applies to all soft water plants.

Duckweed This is another undesirable plant, it develops best in ponds which are protected from winds. In a few days it is capable of covering the whole surface of the pond preventing both light and oxygen from entering the water, resulting in stress and eventual death of the fish. Strong winds will blow the duckweed to the side of the pond from where it can be raked off or scooped out using a fine hand net or moved to one side by a floating beam as previously described when dealing with algae.

A very old and profitable way of getting rid of duckweed is to introduce ducks to the pond (Aylesbury or similar types). At Newhay the author experimented successfully with a stocking density of 50 ducks per hectare (20 per acre), and with no other food source, the ducks grew to table size within three months. The duckweed disappeared and carp benefitted by the growth of zooplankton as the ducks fertilised the water with their droppings. However, the weed may return to the pond after the ducks have been removed in which event the experience should be repeated. The carp in these ponds averaged a weight increase of 15% at the end of a season. The disadvantage of this method is one of additional cost as it is necessary to fence such ponds. Normal fencing will keep the ducks in, but alas will not keep the fox out, and so the odd duck may disappear. This experiment should not be repeated year after year on the same pond, as the water can become over-fertilised. In any case it is not advisable to raise ducks on small ponds or ponds containing C1 or fry.

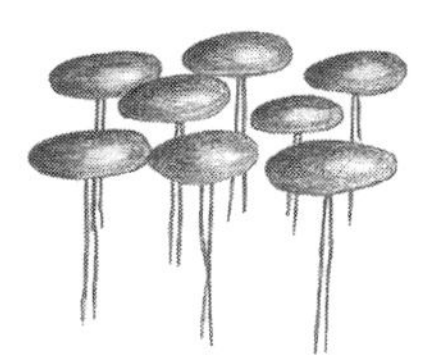

Fig 29 Duckweed (*Lemna* sp) 2–3 mm dia.

There are other underwater plants especially of the *Potamogeton* species which can be controlled using ducks.

'Hard' water plants Other underwater plants are much tougher and are unfortunately difficult to eradicate. One of the more troublesome is water bistort or knotwood (*Polygonum amphibium*); this plant can adapt itself to grow out of

Fig 30 Water bistort (*Polygonum amphibium*)

Fig 31 Common arrowhead (*Sagittaria sagittifolia*) 50 cm high

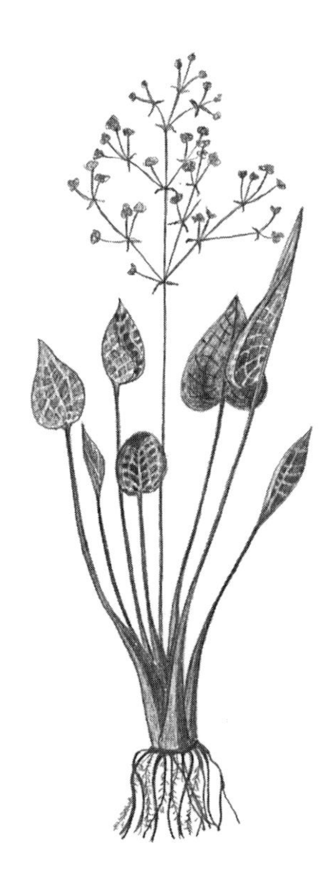

Fig 32 Great water plantain (*Alisma plantago*) up to 80 cm high

water and on land, therefore it can survive in a drained pond. There are a number of species similar to *Polygonum*. Their leaves are fairly large and the stem is hard and of little nutritional value. Fortunately these plants do not spread very rapidly in the pond and are more of a nuisance at netting time than a pest in the actual pond. These include common arrowhead (*Sagittaria sagittifolia*) and the great water plantain (*Alisma plantago*). The best way to control these plants is by using an underwater saw (*Fig 33*), operated by two men. Each blade is twisted and has teeth on both sides. The blades are fastened together to make a long flexible saw. A rope is attached to each end. The saw is operated by rapid movements, each man slowly moving forward. As the saw sinks to the bottom of the pond it

Fig 33 Underwater saw

cuts through the stems of the plants on the bottom of the pond. The cuttings rise to the surface and can be taken out by the raking method previously described. Care must be taken to ensure that all cuttings are recovered from the pond as they readily root and form another plant. The cutting operation should be done prior to seeding and be repeated if necessary. A further method of eliminating these plants is by digging out the roots in winter when the pond has been drained. Chemical treatment is also possible but this should be carried out only in spring time just as the shoots are beginning to show. However, it is not advisable to stock a pond with fish in the same season after chemical treatment.

Emergent plants

These plants are rooted in the pond substrate but the foliage emerges from the water to a height of 2–3 m (6–10 ft). Some emergent species are ornamental *eg* water iris; others, such as water mint are said to have healing properties when taken as a herbal tea. Reeds and bulrushes are of no value to the carp farmer – their roots extract valuable nutrients from both water and soil and may become so dense that they form a barrier between the water and the pond bed. The best way to control these plants is by regular scything.

It is perhaps remiss to end this section without mentioning the gem of all water plants – the water lily. It would be unthinkable to have a garden pond without a water lily. Unfortunately carp ponds need regular draining; the plant can not survive in such conditions where the tuber is exposed to frost. One way to overcome this is to encase the tubers in perforated plastic tube and remove them when necessary.

Fig 34 Bulrush (*Typha latifolia*) 1–2.5 m high

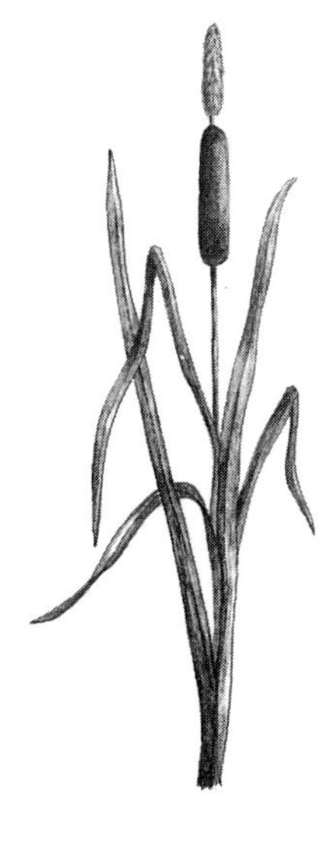

Fig 35 Common reed (*Phragmites australis*) up to 3.5 m high

5 Carp and its food

A comparison with trout

It is interesting to compare the biology and culture of carp with that of the most commonly farmed freshwater fish in Britain – the trout. Although both are fish, they are as contrasting in habit as, say, a pig and a tiger. There are two main differences – the trout is almost solely carnivorous; whilst the carp is omnivorous and consumes virtually anything including animal protein – hence its reputation as the 'pig of the water'.

Trout require a high concentration of oxygen (about 12–14 mg/*l* in summer) and because of the fish's high protein demand (40–60%) the faeces are extremely toxic. Hence, in order to farm trout, a constant flow of highly oxygenated water has to flow through a trout pond to flush out the waste water. As a rough guide – 4.5 million litres of good quality water is required each day to produce 10 tonnes of trout each year. Taking into consideration the cost of trout pellets, about 40% of the total costs must be allocated for feed. To produce trout of an average size of 0.23 kg (½ lb) takes about 8 months. Temperature tolerance is an important factor. Trout stop feeding at low temperatures, feed well at about 10°C, but become distressed and may even die, if kept for any length of time at temperatures of 18°C and above.

Carp on the other hand is a warm water fish. It begins to feed reasonably well at 10°C, best feeding occurs at 18–25°C, but the fish will tolerate temperatures from 40°C down to freezing point when it will hibernate.

To develop from egg to C3 table carp takes about 2½ years (3 summers) after which time an average weight of 0.9–1.1 kg (2–2½ lb) is achieved. To produce 1 tonne of carp in this period requires 0.61–0.81 ha (1½–2 acres) of pond space. Carp require 22–23% animal protein (by dry weight) but the bulk of the diet comes from water plants, their seeds and roots; and supplementary material added by the farmer. Good conversion rates can be achieved, for example, 1 tonne of barley or wheat will produce 1 tonne of carp under favourable conditions. Growth rates are poorer in the absence of additional food items and during unsettled spells of weather.

Zooplankton

The main source of animal protein for carp is zooplankton (mostly small crustacea). One of the most common of these is the water flea (*Daphnia*) of which there are a number of species. The larger forms are particularly favoured as food items by the carp. Two of the best known species are *Daphnia longispina* and *Daphnia pulex*.

The mass production of these small creatures is of the utmost importance to the carp farmer. They occur naturally in almost every open water area – usually introduced at the egg stage by birds. In favourable conditions the egg hatches, always producing a female – if food supplies (phytoplankton, detritus) are plentiful the individual grows rapidly and after a time produces 50–150 live-born 'baby' *Daphnia* without fertilisation. This process is repeated and within a month or so there will be millions of *Daphnia* within the pond – all from a single egg. Sudden changes, such as a drop in temperature or a breakdown in the food chain causes some females to change sex, any males which appear may fertilise the remaining females. The fertilised female produces a single egg and sometimes

Fig 36 Daphnia pulex (water flea)

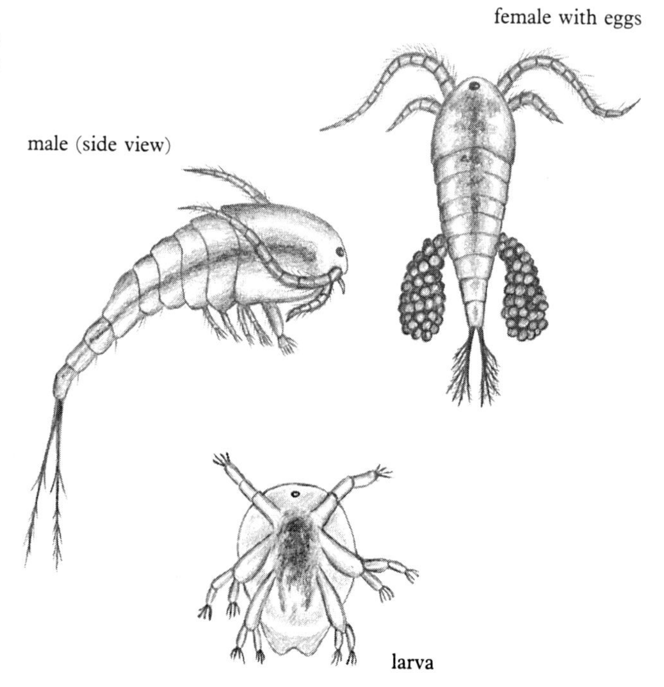

Fig 37 Cyclops sp
1–1.5 mm long

Fig 38 Testing for zooplankton

dies as a result, and the whole *Daphnia* production in the pond collapses. The eggs, which are almost indestructable, lay dormant until conditions become favourable once again.

Other zooplankton species have a similar life cycle. The mass production of zooplankton is almost as important on a successful farm as the production of the carp itself. If the food chain breaks down the animal protein will have to be replaced with pellets, which are expensive. The use of feeding pellets requires more attention to the pond as their use can rapidly affect pond water quality with the build-up of ammonia. To ensure uninterruped zooplankton growth, preparations must begin in winter.

Liming the pond

Liming is carried out when the pond is empty (wintering), best conditions are when the pond bottom is frozen. Addition of lime helps to break up the pond mud and raise the pH value – it also aids the release of nutrients into the water during the next growing season and acts as a disinfectant, helping to destroy disease organisms and parasites which may be over-wintering in the mud. However, liming agents do not destroy the *Daphnia* eggs which may also be present in the substrate. The following substances can be used in carp ponds: powdered limestone (calcium carbonate), quicklime (calcium oxide) and slaked lime (calcium hydroxide). The latter is generally regarded as the most effective. Actual quantities of lime necessary for distribution on the pond bottom varies from site to site, and is influenced by a number of factors, such as the existing pH of the water to be used. Generally speaking, the amount of lime required for an area of approximately 1 hectare would be 250–500 kg. The lime should be applied to the dry pond bottom, but if parts of the pond are impossible to drain, slightly higher doses should be used in these areas.

In some cases ponds are naturally high in lime and additional quantities can have a detrimental effect due

Fig 39 Limed pond

to the excess calcium ions. If joined with phosphate ions calcium produces insoluble calcium phosphate which settles as a deposit on the bottom of the pond.

If the pond can not be emptied completely the lime can be scattered on the water surface. On large ponds this operation is carried out from a boat but care must be taken to ensure that the lime sinks as evenly as possible to the bottom. It is possible to lime a pond whilst the fish are still present, but before doing so the lime must first be dissolved; the lime water solution is then sprayed onto the water. After not less than 2 weeks (generally at the beginning of March) the pond should be fertilised.

Fertilisation

Organic fertiliser

In much the same way as the arable farmer fertilises his fields to improve his yields, the carp farmer fertilises his ponds. The best and by far the cheapest way of doing this is to use organic fertiliser. Cow, sheep and horse manure are all suitable. The manure should be well rotted, this applies particularly to horse manure as it contains more ammonia than the others. As a rough guide, 1 hectare of water surface requires 7.4–14.8 tonnes of manure per season. Two thirds of this should be spread on the dry pond bottom, the remaining third should be distributed around the banks. Over a period of time nutrients continue to enter the pond by the action of rain water on the manured banks. Pig and chicken manure can also be used, but more care should be taken when using either of these as pig feed very often contains copper, this has a tendency to remain in the manure, and copper is detrimental to fish. Chicken manure should be used in much smaller quantities (about 2.5 tonnes per hectare) as it contains about four or five times more phosphate. It is also advisable to store the waste in heaps for at least 12 months before use as most chicken feed today contains antibiotics which in time can have a detrimental effect on the development of zooplankton.

Ponds which cannot be drained are best manured along the inside slope of the pond. It must be emphasised that the beginner should keep to the lower figures quoted in this section, as the effect of fertilisation varies from pond to pond depending on the quality of soil and water.

When manuring of the pond has taken place, it is ready to be filled with water, this should take place between the end of March and the beginning of April. It takes 3–4 weeks of normal spring conditions before a bloom of zooplankton appears, but this process can be accelerated by the addition of further quantities of zooplankton (*Daphnia*). At Newhay a winter supply of *Daphnia* is kept, in ponds which were purpose built for this reason. The ponds were constructed in a very sheltered area to a depth of 4.6 m (15 ft), and are heavily manured each autumn before *Daphnia* is introduced (approximately 500 ml *Daphnia* to each pond). Due to their depth, a temperature of not less than 6°C can be maintained in these ponds and the crustacea continue to breed in these favourable conditions. Every second week a quantity of waste flour (approximately 0.23 kg ½ lb) is sprinkled onto the ponds, which are also aerated. The result of this procedure is a plentiful supply of breeding *Daphnia* at least two weeks earlier than would otherwise be possible.

Inorganic fertiliser

Usually in the middle of the growing season, there is a biological breakdown when phytoplankton formation in the pond can not keep pace with the demands of plankton-eating organisms. The water loses its greenish appearance resulting in a reduction of the zooplankton production. This situation must be remedied, but it is not advisable to apply organic fertiliser at this time of the year as to do so could result in a build up of ammonia. Therefore it is advisable to use inorganic fertiliser which can be applied direct to the surface of a fully stocked pond. A phosphate based compound is suitable for this purpose (nitrogen and potassium fertilisers can also be applied); but the best possible treatment is the use of super phosphate, and one application of 124–185 kg/ha (50–75 kg/per acre) would be sufficient. This should produce a planktonic bloom very quickly and can be repeated after 2–3 weeks if necessary.

Insects and other food in the pond

Many insects start life in the water and as carp pond water becomes rich in minerals and trace elements it provides these insects with an ideal medium in which to lay their eggs. It is, therefore, not unusual on a warm summer evening to observe thousands of mosquitos, sometimes clouds of them, performing their wedding dance over or near carp ponds. The mosquito eggs are laid mainly on the leaves of soft pond plants which start to die off in mid-July, thus providing ideal conditions for the developing mosquito larvae. This is also the time when carp are seeking additional food supplies and so their needs are met by the recently emerged mosquito larvae. These provide the balance of animal protein in the pond and without their presence the zooplankton would be consumed in such quantity that it would be quite unable to reproduce in sufficient numbers to cater for the food requirements of the carp.

Carp seem to favour the mosquito larvae as a food item to such an extent that they temporarily ignore the presence of *Daphnia* enabling these crustacea to maintain a stable population. Natural balance is also encouraged by the carp's habit of digging for 'worms', this activity clouds the water and inhibits algal growth which might otherwise become excessive.

Fig 40 Bloodworm – larva of non-biting midge (*Chironomus plumosus*) 15 mm long

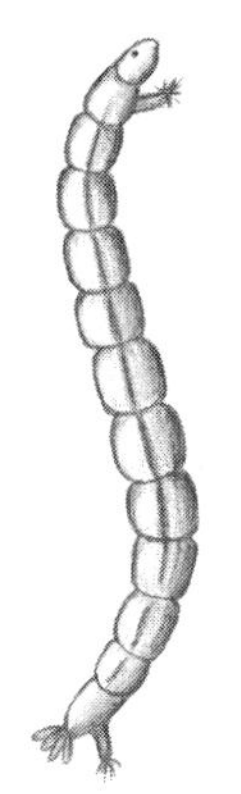

There are a large number of different mosquito types. The very unpleasant stinging mosquito (*Culex pipiens*) is easily recognised in its larval and pupal forms; it hangs itself upside down from the water surface and breathes atmospheric air through a small tube. If disturbed the larvae literally tumbles down to the bottom of the pond. The pupa has a characteristically large head. Another larvae and by far the best as a carp food is that of the blood worm (*Chironomus plomosus*). It grows to a length of approximately 20 mm (¾ in), its high carotene content provides extra colour to the fish, and it is highly valued as a fish feed by fish farmers, ornamental pond keepers and aquarists alike.

Another type of mosquito larvae is known as the glass worm (*Chaoborus* sp). It is almost transparent and moves by a tumbling action through the water. Many more related species are found in the pond, and all

Fig 41 Glassworm – larva of phantom midge (*Chaoborus* sp) 15 mm long

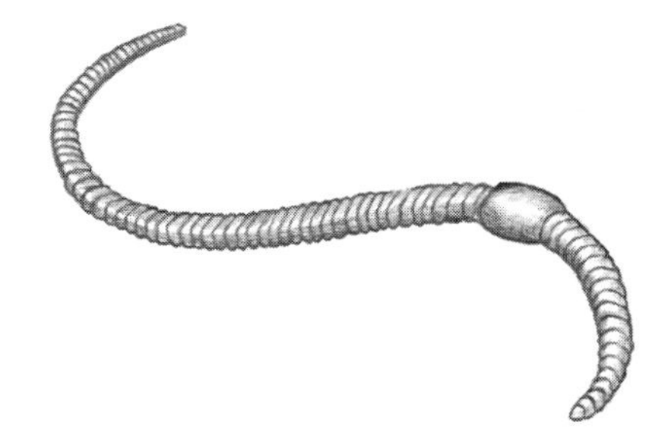

Fig 42 Square-tailed worm (*Eiseniella tetrahedra*) up to 50 mm long

become airborne in the final stage of their life cycle. There are a number of true worm-like creatures that carp will feed on. Two of the most important ones are the square-tailed worm (*Eiseniella tetraheda*) and the sludge worm (*Tubifex tubifex*) – see *Fig 26*. The former is closely related to the earthworm and grows to a size of approximately 50 mm (2 in). It is the only aquatic member of the earthworm family. Its common name aptly describes its appearance, and larger carp love to feed on it. Sludge worms, which grow to about 2.5 cm (1 in) in length are very difficult to detect in the water. They make a burrow, using slime, mud and sand, where they stand head down, the tail protruding halfway out of the mud, and at the slightest vibration will contract into the tube to re-emerge when the danger has passed. These worms are reddish in colour and a large colony of them has the appearance of a piece of red coloured carpet lying in the water, which can quite suddenly disappear at the slightest vibration. In moderation sludge works are a good source of food. If however very large patches of these appear in the pond, it is an indication that the pond conditions generally are deteriorating (as previously described).

As the carp continues to grow and the temperature rises it is not surprising that the carp, in their eagerness for food, will even swallow small snails, snap at insects which accidentally fall into the water, and even attempt to eat some of the smaller fish. However, the carp's most regular and satisfactory method of finding food is by digging in the bottom of the pond, which has resulted in its reputation as pig of the water. Evidence of this can be seen at the bottom of a drained C3 sandy pond which has the appearance of a punctured sheet with thousands of regular sized holes each about 40 mm (1½ in) in diameter and 12 mm (½ in) deep.

Supplementary feeds

Animal protein is an essential part of the carp's diet and must constitute 22–25% of its total food intake. If all goes well during the growing season and the pond is not overstocked with fish, the zooplankton and insect production will be sufficient to ensure maximum growth. The remaining 75% food supply is met by what the carp is able to find in vegetable matter, such as soft water plants, roots and seeds, and what the farmer provides (for the first season up to C1 size, supplementary feeding is given only in special circumstances). Traditionally the most widely used supplementary feeds are cereals. These can be given in different ways; barley, wheat and rye are very suitable as they can be added whole to the pond. The carp waits until the grain is swollen by water before starting to feed on it. Oats should not be given whole as they are too spiky; they can however be given in rolled form. Continental carp farmers often feed lupins, maize or soya, which are also rolled before they are given to the fish. Molasses, brewers' waste and flour mill waste all are good supplementary food. Cooked mashed potatoes are also eagerly taken, but are not of much nutritional value to the carp. It seems that the digestive system of the carp is unable to convert the high starch content of the potatoes properly. Furthermore potatoes quickly turn sour. The addition of potatoes or other quick rotting waste foods must be strictly controlled, and this is done by one of two methods: either providing a fixed platform in the pond on which the food is placed or more conveniently a moveable platform, which can be raised out of the water, filled with food and lowered back into the water to a depth of about 30 cm (12 in) so that water fowl can not easily get to it.

Whenever pellets are given as a supplementary food, the most efficient method of dispensing them is to use a demand feeder. All kinds of pellets can be used, such as rabbit, chicken, pig, carp or trout pellets (but the latter should only be used if the pond is short of zooplankton). Demand feeders are recommended for fish of C2 size and upwards. The mechanics of these feeders is such that the carp has to move a plastic ball on the end of a stick releasing pellets, which it then

Fig 43 Demand feeder

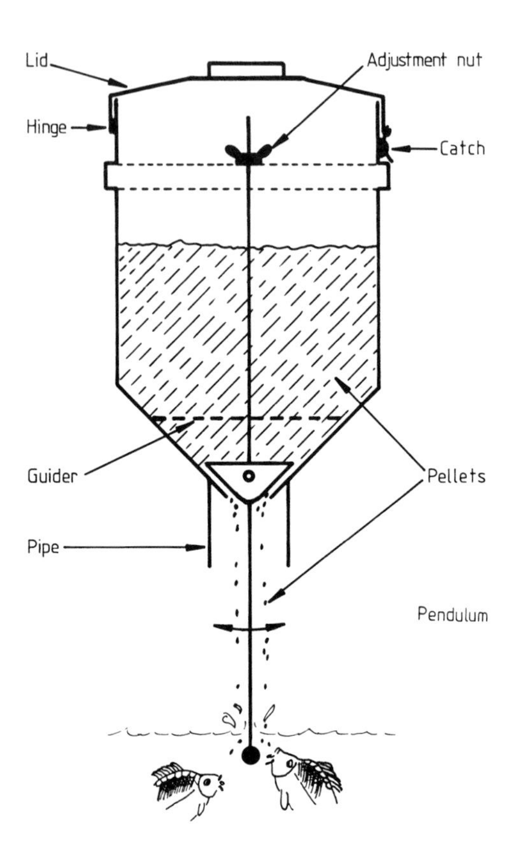

eagerly collects. It is amazing to observe the speed in which the carp learns to operate the feeder. The fish is very selective in its choice of pellets and it will soon be apparent if a combination of pellet types should be mixed together (the carp selects the pellet it prefers and the remaining ones are discarded and fall to the bottom of the pond).

Since the carp is a warm water fish its activity and feeding increase as the temperature rises, hence more food is required in warmer periods. Supplementary feeding, however, should not continue when the temperature reaches 25°C and higher, at these levels the food is poorly digested and quickly contaminates the water. To estimate the amount of supplementary feed needed, a useful reference is the Food Quotient or Food Conversion Ratio. The quotient is calculated by dividing the total weight gain of the fish from one pond into the amount of supplementary feed added. The quotient will vary from pond to pond and with the types of feed used. But after a few years it is possible to calculate the amount of food necessary to increase the weight of the fish by the desired amount. This can be calculated as follows:

Amount of supplementary food = expected growth × food quotient.

Having estimated the total amount of feed to be added during the whole growing season, the carp farmer then subdivides it into monthly portions which are related to the expected water temperatures. The summer months receive a larger proportion of the feed than the spring or autumn months. The fish will be larger in the autumn and so will require more food than at similar temperatures in the spring. A typical schedule for supplementary feeding is:

Month	*% of annual supplementary feed*
May	10
June	15
July	25
August	30
September	20

Supplementary feed should be given every 2–3 days during the summer and less frequently during spring and autumn. Supplementary feed is usually only given to C2 fish.

6 Carp breeding

Natural breeding

The carp is a warm water fish and so it is not surprising that breeding takes place in the summer months. It usually begins in late May or early June, but can be later than this if the water fails to reach and maintain a temperature of 18°C (64°F), for a reasonable period of time. The ovaries slowly begin to ripen after a hibernation, and by the end of May they are fully developed. If breeding conditions are unfavourable there will be an automatic halt by both sexes to the breeding process, and this can last for a number of weeks. As soon as conditions become favourable a flow of hormones from the pituitary gland

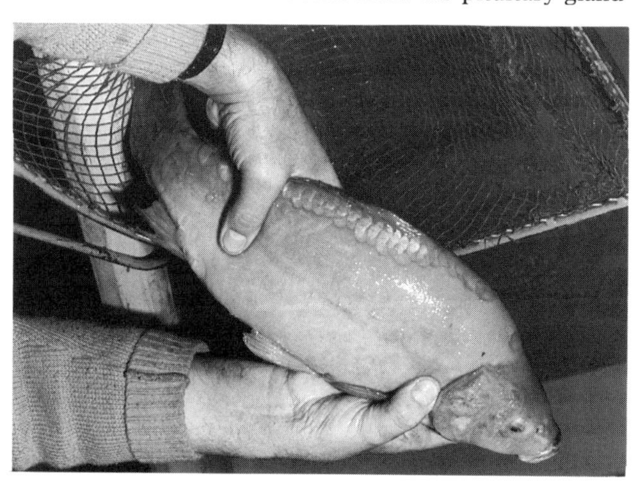

Fig 44 Mirror carp (*Cyprinus carpio*)

sets the 'wedding play' into motion.

The males develop pimples on the gills and wild chasing of the females begins. The males attempt to rub their gills along the soft body areas of the female, and this action triggers off ovulation and with it release of the eggs. These are immediately fertilised and vigorously splashed about. As the eggs move slowly through the water they become adhesive, and attach themselves to foliage or to roots. Eggs which fail to attach themselves to something will fall to the bottom of the pond and perish. The courtship starts early in the morning, about 6 am–7 am and finishes about noon. Some of the carp are then so exhausted that they can be lifted out of the water by hand. There are now eggs everywhere sticking on plants and roots, they are very small, about the size of a pin head. A female of 4.5 kg (10 lb) in size is capable of carrying ½–¾ of a million eggs. Twelve hours or so later the fertilised eggs become crystal clear and the unfertilised eggs become milky in appearance. For the next three weeks the temperature must not fall below 13°C (55°F). The eggs take 6–10 days to hatch, depending on the water temperature. In a further 12–14 days the fry passes from its larval form into what resembles a small carp.

Fig 45 Female mirror carp ready for breeding

Only after this stage is the fish able to withstand lower temperatures, and for short periods only. Only the C1 size carp is able to cope with all the fluctuations of temperature.

The number of fertilised eggs will depend to a great extent on the condition and size of the pond; 30–40% is considered to be an acceptable percentage. A large number of eggs start to develop normally and then fail, especially when a number of eggs are clotted together. A large number of 'enemies' (from very small insects to wild fowl) destroy eggs and fry, so that at the end of the season when the carp reach fingerling size, only about 1% of the eggs from one female will have survived (see *Predators*).

Breeding in Dubish ponds

It can be seen from the previous section that natural breeding is somewhat hit or miss, and carp can only maintain a population because of the very large number of eggs a female carp can carry (up to 2 million). Even in countries blessed with a moderate climate, there are years when carp fail to spawn, the fertilised eggs do not develop, or the larvae die mainly because the temperature falls at a crucial time to below 13°C (55°F).

Not surprisingly, carp farmers, even in the early days, were looking for better methods of breeding. These came from a rather unexpected source. Thomas Dubish was born in 1813 a son of a Danube Fishery Master. As a boy he spent much of his free time on the banks of the Danube and the surrounding countryside. In his keen observation he noticed that the carp were spawning at the time of year when the Danube was in flood, due to melting snow from the Alps. Later in his life he became a Fishery Master himself and his work took him to a different part of Austria. He never forgot his observation as a boy, and dug a small pond in a sunny and sheltered spot. This pond could be completely drained. He seeded it with coarse grass and then flooded it with well water before leaving it to stand for sufficient time to allow the water temperature to reach the desired level. He then introduced a breeding trio of near ready carp, and they spawned. He lowered the water level to remove the breeding fish, and then filled the pond up again with temperate water. This was repeated again and again with great success. Unfortunately he was unable to read or write, so it was only many years later that one of his apprentices committed the details of his experiment to paper.

Later still, a man named Hofer altered the Dubish design slightly, as he was operating a carp farm in a higher and therefore cooler region (see *Fig 46* showing the differences in design). The slope at the bottom of the Hofer pond is much more pronounced. Hofer tried to achieve slower cooling of the water by increasing the depth on one side of the pond more than that of the Dubish pond. Therefore the Hofer design appears to be the better choice for use in the British Isles. To give extra protection from wind a conifer hedge can be planted on the opposite side to the sun, or artificial wind breaks can be created. These breeding ponds should be constructed in such a manner that they can be drained and kept dry. They should then be seeded with hard grass seed such as rye grass.

Fig 46 Section diagrams of Dubish and Hofer ponds

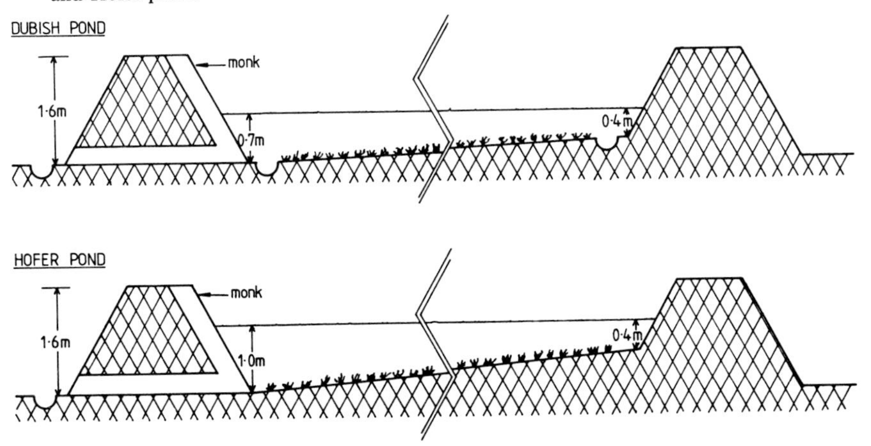

Selecting brood stock

Selecting carp for breeding begins in mid-April after the fish have come out of hibernation and are searching for food. Only the very best fish should be taken, females not less than 5 years old and males not less than 3 years old. Fish can be easily sexed at this time of the year as in the females the eggs are well developed and have started moving downwards and the lower part of the body should feel somewhat spongy to touch.

Male fish are much thinner than females. The carp farmer should ensure that the fish selected for breeding purposes have no deformities of the spine, fins or tail.

If the temperature is 10°C (50°F) or over, which it should be in April, a sperm test can be made. The best way of doing this is to hold the fish in a wet blanket to get a good grip, then slide the hand towards the tail applying slight pressure with the thumb and fingers. A slight sperm flow should then be visible, similar in appearance and consistency to cream. Thin and bluish looking sperm should be rejected as unsuitable. It should further be determined that the breeding fish are free from parasites, mainly fish lice and leeches. Fish can be treated for these conditions (see *Parasites*), but should not be used for breeding purposes afterwards.

Whenever possible the brood fish should be kept separately in smaller ponds so that uncontrolled breeding cannot take place. Careful study of long range weather forecasts should be made before the carp farmer chooses the right time to fill the ponds with water ready for breeding. After this has been done, the water should stand for 3–4 days, and when the water temperature reaches 18°C (64°F) or higher, the breeding fish (two males and one female, or in larger ponds two such trios) can be introduced.

It may even be advisable to have two breeding ponds side by side, the reason being that a fairly high percentage of fish (10–20%) are not suitable for breeding (*ie* infertile). If for some reason the trio have failed to spawn within 4–5 days, the fish should be taken out and the pond emptied and refilled before another trio is introduced. Great care should be taken when transporting breeding fish. The best method is to wrap the fish in a wet towel. Usually, spawning takes place soon after introduction, often in the early hours of the next morning or the morning after that. Delayed spawning is often due to the unfamiliar surroundings of the breeding pond (particularly if the fish have been brought from a much larger pond). This is another reason for holding fish in a smaller pond prior to breeding. After the fish have finished spawning (around midday) the carp are in a state of exhaustion and can be easily netted out by hand. The water level can be lowered by half to make this process easier. The water should then be replaced, making sure that the incoming water is of the correct temperature. Most of the eggs will now be firmly attached to the grass. The fertilised ones will have started the development process, and the unfertilised eggs (which die within 2–3 minutes) become mouldy and furry very quickly, and can be easily seen for a time.

The sperm cells also die off fairly quickly in these conditions, but sperm can be kept for a considerable time if placed in a refrigerator.

Hatching of the larvae depends entirely on the temperature. At an even 23°C (75°F) the eggs will hatch in 3 days, but in a breeding pond the same process usually take 6–10 days. If temperatures remain between 13–30°C the eggs will develop normally. However, the wise carp farmer should be prepared for conditions to be unfavourable (a sudden drop in air temperature will result in dangerously low water temperatures), and to overcome this, the pond should be protected by placing a plastic sheet over the water surface. An alternative is the use of electric or bottled gas heaters. On occasions the pond water may become too warm. If this should happen, clear cold water should be added to the pond and the water aerated.

After the fish have hatched they will rest for a time on the grass and for the first 2 days of life will draw nourishment from a yolk sac. During this time the larvae make repeated efforts to swim to the surface in order to fill their tiny swim-bladders with atmospheric air. The first attempts at swimming produce awkward jerky movements but soon the fish begin to look for food. It is at this stage that they should be transferred to a well prepared fry pond. A very fine hand net and a bucket are used, and care must be taken to ensure that the water temperature in the bucket is equal to that of the fry pond which should be rich in micro-organisms such as rotifers and protozoa *etc*. It can also contain *Daphnia*, but not cyclops as at this stage the crustacean could attack the larvae. At the time of transfer the larvae are 5–7 mm in length, but they grow rapidly in a well prepared fry pond and within 10–14 days, depending on the temperature, they become small carp with all the features of their parents.

In the fry pond the food supply soon becomes exhausted and the young fish have to be moved again. This time to a much larger pond which is rich in zooplankton. They remain in this pond until the end of the season, by which time they should be handsome fish of 10–12 cm in size. Removal from the fry pond is usually achieved using small traps, which have been suitably baited to attract the fish.

7 Induced breeding

After the Second World War great changes took place in almost every industry and carp farming was no exception. Fertility drugs began to be available, for use on humans and animals alike. It was discovered that a small gland, the pituitary, plays many roles in the behaviour of many creatures. This gland has a number of functions, and one of them controls sexual behaviour. Experiments with sea fish, not previously bred in captivity, produced some good results.

In trout farming induced breeding is not necessary as trout ovulate for a whole month, during this period the eggs and sperm can be easily extracted. Carp, however, ovulate for only 5–8 hours. In order to predict the ovulation period it is necessary to inject the carp with pituitary material. In other words, the loosening of the eggs from the ovaries can be predicted and the eggs can then be extracted. The use of pituitary material in induced breeding was a great step forward and has many advantages. The whole breeding process can be taken into the laboratory, and carp can be produced all the year round. Instead of about 20% hatchability (as in Dubish or Hofer ponds) the fertility of the eggs with induced breeding is about 70–80%. However, as always, there are certain disadvantages. Induced breeding is more expensive and needs more experienced human assistance.

In large hatcheries the breeding fish are kept over

winter at average feeding temperatures of approximately 15°C.

The females should be separated from the males as there is always the possibility that a female will ovulate early (minimum ovulation temperature is 18°C).

If it is not possible to keep the fish in a laboratory they will be wintered in an outside pond and if possible the females should be separated from the males. There is a well known formula for breeding the fish brought in from wintering ponds. It is called the One Thousand Centigrade Days system, abbreviated as C/D. This formula is for use only for fish which have come from a hibernating temperature of approximately 4°C and it allows the female the required time to develop the ovaries until the eggs are ready for ovulation.

For example

The breeding fish are brought into the laboratory on the 31st December. The water temperature at this time both in the pond and the hatchery being 4°C. The next day the water temperature should be raised by 3°C. We then have a formula of 7 Centigrade Days (C/D). On the second day the water temperature is raised again by 3°C and at the end of the day the formula will be 10 Centigrade Days (C/D). This process continues until the breeding temperature of the water has been achieved. This is between 23 and 25°C.

Care should be taken when raising the water temperature not to exceed 4°C per day and when lowering the water temperature this should not exceed 2°C per day. For the purposes of this example we are aiming at a temperature of 25°C and this will be reached in 7 days time. We refer to that figure now as the first 25 Centigrade Days. The water temperature is now 25°C and will remain at this temperature as we add 25 C/D each day on paper, thus at the end of the day our records will show that we have 25 C/D from the first seven days, plus 25 C/D added on the 8th day.

After a further 2 days our figure will be 100 C/D. After a further 4 days the figure will be 200 C/D. Therefore in 46 days, calculated from the first day the fish were brought out of the pond our written records will show *1000 Centigrade Days* and this is the approximate time that the pituitary injection should be given to both the male and the female fish.

For breeding carp later in the season when the outside pond temperature has been rising for some time, or for carp which have been kept in the laboratory at a higher temperature the preconditioning period is shorter due to the natural development of the ovaries as the water temperature increases. The fish are ready for breeding when by applying slight pressure towards the vent of the male a flow of sperm is released and in turn the females should look quite heavy and feel spongy to the touch. It is now time to prepare to inject the fish. It can not be overemphasised that from now on cleanliness is of paramount importance and every piece of equipment must be tested and checked, Zuger jars kept running (see *Figs 53 and 54*), the temperature regulated and all chemicals prepared in advance.

Equipment required for induced breeding

2 – buckets
2 – 5 *l* bowls
3 – 2.5 *l* bowls
3 – 5 *l* plastic containers
1 – flask (1,000 ml)
1 – homogeniser
bottles of injection caps
syringes and needles (1 ml)
syringes and needles (5 ml)
balances, 1–3 kg and 1–12 kg
measuring cylinder (1,000 ml)
fridge (or container with ice cubes)
a number of large goose feathers
pituitary harmone
fertilising solution

tanning solution
MS 222
beakers (1000 ml)
beakers (250 ml)
distilled water
towels.

Injecting the fish

The pituitary gland material can be obtained either in powder form or the complete gland (one gland is about the size of a pin head and weighs about 3 mg). Whether powder or gland, the pituitary material must be liquidised as follows: dissolve 6 g of sodium chloride (common salt) in 1 *l* of water, then dissolve 3 mg of pituitary in 0.5 ml of this salt water. The amount to be injected into the female carp is calculated from the weight of the fish – 3 mg pituitary solution for each kilogram of fish weight. 20% of this dose is injected 12 hours before the main injection. From a timing point of view, it is best to give the first injection between 8 am and 10 am. The main injection is then due between 8 pm and 10 pm, and the fish will be ready for stripping between 8 am and 10 am the following day. The male receives just one injection, and the strength should be 1.5 mg pituitary dissolved in 0.5 ml

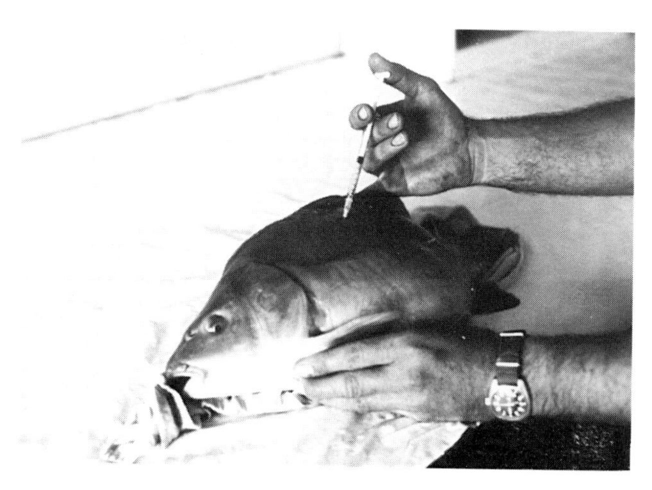

Fig 47 Injecting a brood carp

of salt solution for 1 kg of carp weight. This injection takes place at the same time as the second female injection, *ie* 12 hours before stripping.

Throughout the injection period the fish may show some signs of discomfort, especially the females. Colour changes may be evident, due to the rapid hormone changes the fish are going through, or sometimes the injected hormones clash with those released naturally in the fish causing further distress. To keep stress to a minimum supplementary feeding should cease for a time before stripping occurs.

At stripping time (24 hours after the female's first injection) the males are dealt with first, which is usually not difficult. The fish should be placed in a damp towel to absorb surplus water with a plastic bowl beneath the vent, then by gentle massaging towards the vent, sperm should begin to flow freely. Avoid any water mixing with the sperm, and take care that the stripping is done as gently as possible so that no blood vessels burst in the fish. The sperm container should then be set aside in either a refrigerator or placed in a bowl containing some ice cubes, to maintain a temperature of between 3°C and 4°C.

The females are handled in the same way. The eggs should flow out by applying only very gentle pressure,

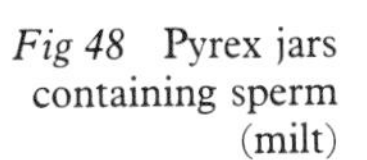

Fig 48 Pyrex jars containing sperm (milt)

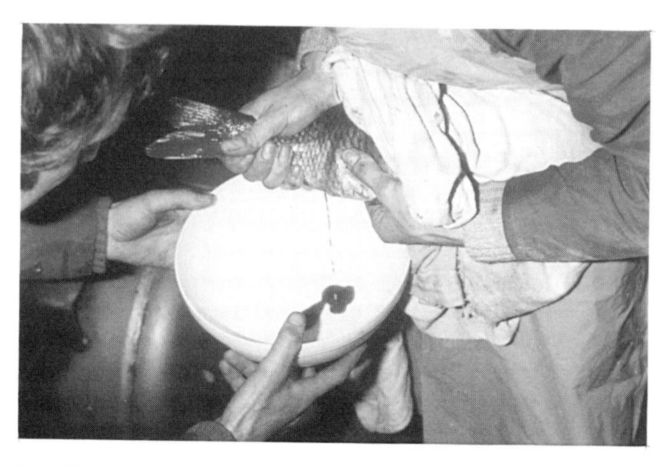

Fig 49 Stripping a female

but if none appear, the fish should be returned to the tank and a further attempt made about half an hour later. If after this there are still no eggs, no further attempt should be made.

All being well, the eggs have now been extracted into a fairly large bowl. Again it must be emphasised that no water must come into contact with the eggs, as this results in clotting and the eggs being spoiled. The quantity of eggs can be measured by marking the outside of the container and from this the amount of sperm required for a given quantity of eggs can be calculated. The ratio is 100 : 1, *ie* 1000 ml of eggs need 10 ml of sperm.

Using a large goose feather the eggs should be very carefully stirred, and as soon as the correct quantity of sperm is to hand, it should be very carefully mixed into the eggs with a constant gentle stirring action. To make the sperm more active a salt/urea solution is added. This solution, which is prepared beforehand, consists of 1 *l* of water (this can be tap water, which has been standing for 2 days and has been aerated so that all additives have evaporated), 3 g of common salt and 3 g of urea. Small quantities of this solution should be added to the egg/sperm mixture. The total amount of fertilising solution added should be roughly equal to

Fig 50 Stirring eggs with a goose feather

Fig 51 Eggs beginning to swell

the volume of the egg/sperm mixture.

As fertilisation takes place, the eggs begin to swell. The salt/urea solution should be added slowly in very small quantities; adding too much too quickly is harmful. Within the next half hour the swelling of the eggs will be apparent, and within one hour they will have swollen to five times their original size. Stirring may be continued for a further 10–15 minutes only, but after this time fertilisation cannot be any further

Fig 52 Tanning solution is added to remove the adhesive properties of the eggs

improved. The tanning solution should now be to hand to remove all the adhesive properties of the eggs. To make this tanning solution: take 1 *l* of water and add 5 g of tannic acid. Mix well together, then add more prepared water to make up to 10 *l*. As an example for the use of the tanning solution take 2.5 *l* of swollen eggs and place them in a large bowl (at least 5 *l* size) pour 0.25 *l* of tanning solution over the eggs and stir quickly adding at least a further 1 *l* of prepared water to the mixture, stirring constantly; then drain off as much liquid as possible. Repeat the whole process twice more. (It is essential that all the liquid should be at a temperature of 23°C). The eggs should then be transferred immediately to Zuger jars, where the water flow is adjusted in such a manner that the eggs are suspended without flowing out of the jar. The whole tanning process must be carried out very swiftly, and is one of the most difficult parts of induced breeding. It is best carried out by two people and should be rehearsed a number of times. If the eggs in the Zuger jar show signs of clotting, another tanning bath may be given, but this must be carried out very quickly; if the eggs are immersed for too long the tannic acid will attack the delicate egg shell.

Fig 53 Standard Zuger jar

Fig 54 Zuger jar containing 150,000 eggs

The eggs are now incubating and their progress can be checked at any time. One standard Zuger jar will hold 2.5 *l* of swollen eggs (1 *l* of eggs equals roughly 150,000 eggs). At a temperature of 23°C the eggs will hatch in approximately 3 days.

Carp and especially koi grow very unevenly at the fry stage, and some reach a size which allows them to swallow smaller fry. This unfortunate stage begins when they are about 15 mm in length. Once these larger fish have succeeded in swallowing fry they develop an appetite for it and grow even faster. This habit develops more quickly if the larvae have been fed on egg yolk rather than *Artemia salina*. This practice will persist for 3–4 weeks, after this time, the smaller fish will have grown to a safer size and can not be so easily consumed. Losses of fry in this way can be considerable, and therefore the fast growing fish should be segregated from the others. A simple way of doing this is to use nets of the appropriate sizes, whereby the smaller fish can escape and the larger fish are trapped. Cannibalism only happens where fry are raised in indoor tanks or containers; in these conditions, unlike open ponds, they have no means of escape.

Summary of induced breeding

It can be clearly seen that the development of induced breeding techniques has brought a number of advantages to the carp farming industry.

- Fry can be produced all the year round, which is especially valuable to breeders seeking to break into Third World markets.
- As natural spawning of carp does not occur until late May/early June, induced breeding can take place in the laboratory earlier in the year. For example, fry hatched in mid-February, and reared in troughs in the hatchery until mid-May, can then be placed in outside ponds, which by this time will be full of natural zooplankton. The result will be that these fish will have grown to full C1 size

(7.5–10.0 cm/3–4 in) by the time natural spawning would just be taking place, and the breeder will therefore have gained a full season, as these fish are comparable in size with fish he would otherwise have had to overwinter. This is of even greater advantage to koi breeders; at least 90% of koi are sold at this time of the year (mid-May) at C1 size. Induced breeding will give the breeder a quick return on capital, and will also save pond space, which in turn can be used for either wintering larger specimens or taken as an opportunity to recondition the pond.

- It must be remembered, that in order to succeed using natural breeding methods, the breeder is at the mercy of the weather, and if at any time, between the carp spawning and the fry reaching the age of 2–3 weeks, the pond temperature should fall below 13°C, the whole effort of producing carp is likely to have been in vain. Even in the best known European carp producing districts it is accepted, that one year in five will be a failure. It is only fair to comment that because of the inconsistencies of the British climate, the success rate in the British Isles will be even lower. Under controlled temperatures in a hatchery, failures are rare. In fact, in a well ordered hatchery 80–90% hatchability of the eggs is not uncommon. If problems should arise, it will be more likely to be the result of malfunctions in the system, or carelessness on the part of the farmer.
- Finally it must be mentioned that running a carp hatchery is rather expensive and time consuming during the breeding period. It may therefore not be financially viable for the small-scale farmer.

Anaesthetics Young breeding fish of 5–7 years old can be handled easily enough during the stripping period without anaesthetics, but larger fish are very strong and holding a large female in the correct position for stripping can present a problem. There is also the danger of dropping the fish which could result in all the eggs bursting out from the body, and even the loss of the fish. To avoid these problems and unnecessary suffering to the fish, the use of anaesthetics is recommended.

The best anaesthetic for this purpose is MS 222 (available from veterinary surgeons). The suggested quantity is 2 g/10 *l* water in a container large enough to allow the fish to be lowered into it. The fish should be carefully watched as it slowly quietens down. The time to remove it from the bath is when it no longer responds to the touch, the gills however should still move freely. Too deep an anaesthetic can be harmful to the fish. An added advantage in anaesthetising the fish is that the vent of the female can be stitched, thus preventing the accidental loss of eggs. The stitches can be easily removed after stripping.

After the stripping process has been completed the fish should be placed in well oxygenated water, where for a few minutes it will lay on one side and then swim quite happily.

Incubating the eggs

The eggs are now incubating in the Zuger jars and turning slowly suspended by the water flow, which comes from below. As the eggs at this stage are very delicate, the water supply must be reduced to a minimum flow (a rate of approximately 0.5 *l* per minute, adjusted according to visible conditions in the jar). Too strong a flow in the first 10 hours of incubation will result in deformities of the embryo, which would be clearly visible in the growing fish. After 10 hours the water flow should be increased to about 1 *l* per minute as by this time the developing eggs require more oxygen; again, some adjustments may be necessary. At this stage it is easy to see the eggs which have not been fertilised as they appear milky. It is advisable to remove them, and this is best done by using a long glass tube. By closing the top end

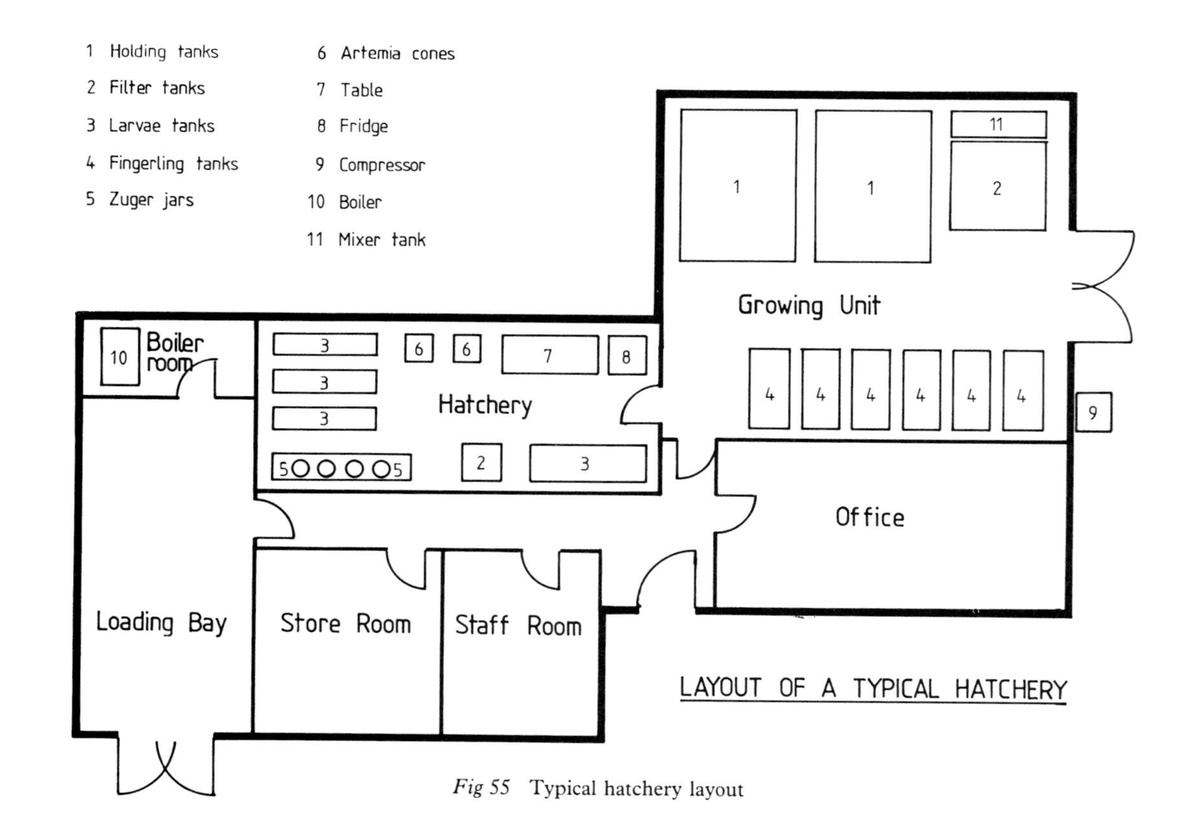

Fig 55 Typical hatchery layout

with the thumb and directing the tube close to the unfertilised eggs, they are sucked in when the thumb is released from the top of the tube. This method is successful, but requires practise. If unfertilised eggs remain in the jar they do not pose any immediate hygiene risk to the healthy eggs. The former become furry which increases their buoyancy and eventually lifts them over the edge of the Zuger jar and into the trough below, from where they can be easily removed.

After about 3 days the embryos are showing signs of increased activity inside the egg. Eventually the eggs break open and the larvae hatch, usually tail first. At this stage they are exhausted and attach themselves to the wall of the Zuger jar, where they remain for about 2 hours, gaining energy. Suddenly they try to propel themselves to the surface of the water in an attempt to gulp down atmospheric air, to fill their tiny swim-

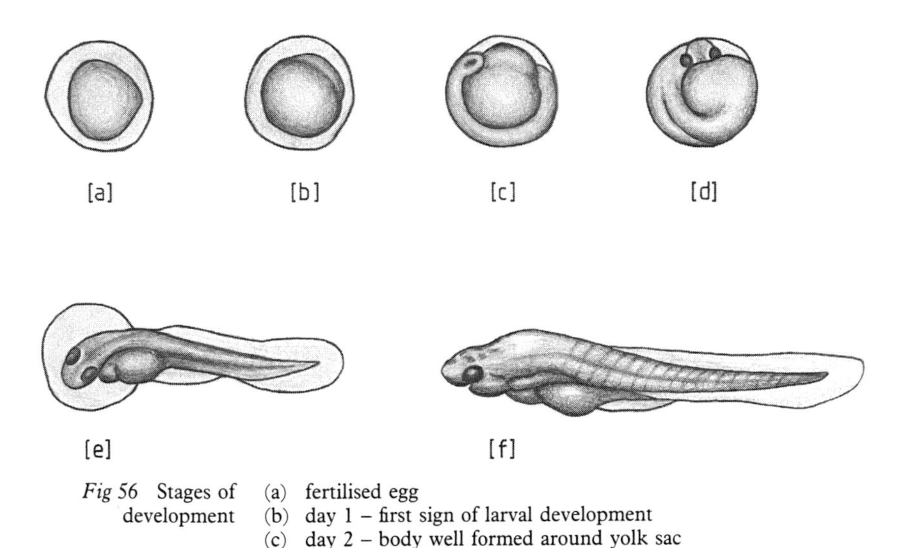

Fig 56 Stages of development

(a) fertilised egg
(b) day 1 – first sign of larval development
(c) day 2 – body well formed around yolk sac
(d) day 3 – head and eyes clearly visible, heart beating
(e) day 3 + – hatching fry (actual timing depends on water temperature)
(f) hatched fry with yolk sac

bladders. As they attempt to do this, the flow of water will carry them over the top of the Zuger jar and into the trough. The larvae do not suffer any adverse effects from the relatively large drop. After entering the trough in this way they alternate with periods of resting and taking nourishment from their yolk sac until they become free swimming. This stage of development covers a period of about 2 days and by this time the trough is teeming with jerky moving larvae. The yolk sac will continue to provide for their food requirements for a further 3–4 days depending on the temperature, but after that time it will be necessary to provide food for them.

Feeding the larvae

Two methods of feeding larvae have been developed; the first method uses hard boiled egg yolk, but should not be continued for longer than a week. The very hard boiled egg yolk should be spread thinly and evenly over the water surface, as any unconsumed egg yolk will quickly contaminate the water. During this process the water flow through the trough should be increased and sediment from the bottom regularly removed. At the end of the feeding period the larvae should be removed and placed in a well prepared fry pond containing an abundant supply of food micro-organisms such as rotifers and protozoa. The transfer can only take place when the pond water temperature is around 20°C, meanwhile the water temperature in the trough should be gradually lowered by not more than 1°C per day, until it corresponds to that of the pond. It could be argued that it would be easier to introduce the rotifers from the pond into the trough, but this is not advisable as the food material is almost certain to be contaminated with harmful bacteria and parasites, which will spread quickly in the trough due to the high density of fish larvae there.

A second feeding method is used when breeding takes place out of season or when the young fish have to be kept for a long period in the hatchery. In such cases the fish larvae are fed on newly hatched brine shrimp (*Artemia salina*) a very small seawater crustacean. The fully grown shrimp is about 20 mm long but the eggs are minute as are the newly hatched larvae which makes them an ideal food for the young carp fry. The eggs are supplied from the salt lakes of North America (mainly in Utah) and kept dry, in a sealed container. The eggs remain viable for years. In order to hatch these eggs it is necessary to create sea-like conditions. A hatching container (of 20–25 *l* capacity – 4–5 gallons) should be built from marine plywood if possible, and given a coating or two of resin (see *Fig* 57). It should be three quarters filled with tap water (tap water has usually a pH value of 7.0 – a little sodium carbonate should be added to bring the pH value up to 8.0–8.6). This well aerated water should be left to stand for 2 days before adding sea salt or non iodised salt (approximately 15 g/*l* water) at a temperature of 22–25°C. Two level teaspoons of *Artemia* eggs should then be added to the solution. With strong aeration and good light, they will hatch in about 2 days, the aerator should then be switched off for 10–15 minutes. The hatched larvae will then sink to the bottom, and the shells, which are lighter, will

Fig 57 Cones for hatching *Artemia* eggs

float and can be removed using a small net. The *Artemia* larvae should then be siphoned from the bottom again through a very fine net, and then fed to the fish. The larvae will survive in fresh water for about 3 hours. The fry should be fed every 3–4 hours and at least four or five times a day; and it is recommended that two *Artemia* 'cones' be used to ensure a continuous supply of food to the fish (*Fig 57*).

Feeding with *Artemia* should continue for 2 weeks by

Fig 58 Growing troughs with electronic feeders

Fig 59 Large fry in troughs

Fig 60 Recycling filter unit for growing troughs

which time the larvae will have developed into small fish, which average 1 cm in length. The fry are now ready to be transferred to a well prepared growing pond with a good supply of zooplankton food; where they will remain until the end of the growing season.

Alternatively, the fish can be transferred into the growing troughs, where they can be trained to take a specially prepared food (trout starter with added trace elements).

In a normal season, and if not over stocked, the fish in the growing pond should reach 8–16 cm. It is somewhat characteristic of the growing carp that they do not all grow at the same rate. Additional feeding should only take place when zooplankton supplies are exhausted. Losses of 10–20% of C1 during the growing season should be considered as normal (see sections on diseases, parasites and predators). The stocking density for 0.5 ha of well prepared pond is approximately 15,000 fish of C1 size; if at the end of the season 12,400 of these remain and are harvested, the project can be regarded as successful.

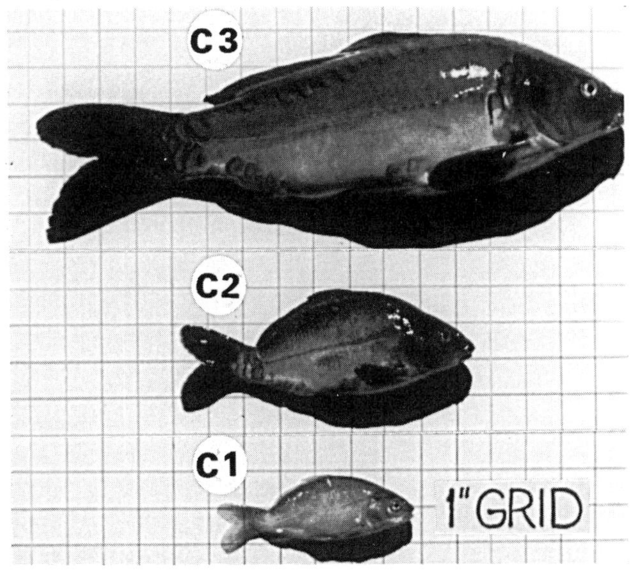

Fig 61 Sizes of C1, C2 and C3 carp

8 Stocking the wintering pond

To winter carp successfully, the fish must be in good condition before the onset of winter. C1 fish continue to feed in temperatures as low as 5–6°C, so if zooplankton are not evident in the pond a small amount of protein food should be added (2–3 mm trout crumbs – 50% protein). A number of C1 should be trapped for examination; they should be 7–15 cm in size and look plump. An undernourished fish can be detected by observing the eyes which will have a sunken appearance. The gills should be evenly dark in colour with no sign of whitish looking fungus. Any parasites particularly leeches or fish lice should be removed (see section on treatment for parasites).

It is probably best to transfer fish to a freshly prepared wintering pond but if this is not possible, the water in the growing pond should be adjusted to a depth of 1.40 m (see section on wintering ponds). C2 and C3 fish cease feeding earlier than the C1 (8–10°C). For these older fish, the most obvious sign that all feeding has stopped prior to hibernation, is that the water will be noticeably clearer than in the growing season, indicating that the carp has stopped digging for food. The number of carp to be placed in a wintering pond is largely a matter of experience, and depends on the situation of the pond and the water quality. As a rough guide – a 1.40 m deep pond of 0.4 ha in size with a continuous water flow of about 900 *l* per hour will accommodate 4–5 tonnes of wintering carp (C2 and C3). For C1 size – about 1000 fish should be introduced for every cubic metre (30 cu. ft) of water; in ponds of 0.1–0.2 ha (¼–½ acre) in size. The ideal situation for the carp farmer is for a long warm autumn followed by an early spring. A prolonged cold spell between does not present serious problems, and is in fact preferable to widely fluctuating temperatures. A warm period even in mid-winter will be enough to mobilise the carp, and even in a semi-conscious state, they will start to look for food. This moving around involves the use of energy which is difficult to replace during the winter. In C2 and C3 wintering ponds such movement can be easily detected by the pond water becoming cloudy. In C1 ponds however such movements will only be apparent by checking the temperature of the water. If the temperature reaches 6°C and is maintained for a number of days and nights, the C1 carp will take some food, so a small quantity of trout pellets of the appropriate size should be given at various points around the pond during these periods.

If the ponds should freeze over for a couple of days no harm will come to the fish, as grasses and water plants growing around the pond, will allow underwater gases to escape. But frost can be harmful to the fish, when snow and sleet accumulate on the ice and prevent daylight from penetrating the pond. Ice should not be broken by the use of hammers or axes, a far better method is to remove squares of ice using a saw, snow should always be removed from the ice if possible.

Stocking density for the growing season

As spring approaches the temperature in the pond will rise, and the carp should then be transferred back into growing ponds. In order to calculate the number of fish with which to stock the pond, two factors should be considered:

- the quality of the pond;
- the size of the fish.

In a normal farming cycle only C1 and C2 size carp need to be relocated as C3 size carp should either have been sent to the market or sold for angling purposes. A realistic initial stocking density for the growing pond is:

10,000–12,000 C1 size per hectare (4000–5000 per acre);
1,000–1,300 C2 size per hectare (400–600 per acre).

At the first glance these figures may suggest that there are too few fish for the given areas of water, particularly when compared with stocking densities for trout, but this reflects the different economic considerations involved in the farming of the two species. With a stocking density of 1000 C2 size fish per hectare (400 per acre) the growth rate should be 0.9 kg (2 lb) per annum; producing a larger profit margin (even if C2 fish need to be bought in) than that from 1 hectare (1 acre) of prime arable land. Potential profits increase still further when more than one pond is available. With a second pond the farmer can produce his own C2 fish (and hence avoid buying in) and there is usually a steady demand for any surplus fish produced.

Even with the less optimistic view that every third or fourth year is likely to bring a poor growing season, the extra income from even a couple of ponds can be quite substantial, often from an area where little or nothing was earned before.

9 Semi-intensive carp farming

It is possible to farm successfully and achieve good growth rates with a higher than average stocking density, by installing a filter aeration unit in the pond (see *Fig 62*). One such unit is sufficient for a 0.3 ha (¾ acre) size pond. The filter unit is installed as follows: a hole about 7.5 m (25 ft) square and 0.6 m (2 ft) deep is dug in the middle of the pond and filled with hard gravel (2.5 cm/1 in) size. The filter unit is then embedded in the gravel. An air line connects a ¾ hp oil free compressor to a diffuser in the cone of the aerator. The pond is then filled with water and the compressed air is forced out of the diffuser as thousands of small bubbles. The air passing through the cone creates a vacuum inside the unit, which, by the action of the bubbles being forced through the narrow opening of the filter aerator, draws water through the four slots at the base of the filter unit and through the gravel bed. Bacteria attach themselves to the gravel and feed on the impurities carried in the water, thereby purifying the water. The system can be likened to a small sewerage plant within the pond, which oxygenates the water at the same time. The unit becomes fully active after it has been running for one week; it can then be controlled by means of a time switch.

A pond containing such a unit can be stocked with twice as many fish as a naturally maintained pond, but

Fig 62 Filter-aerator unit

Fig 63 Installing the filter-aerator

requires more attention. The water quality needs to be checked more frequently and fish need more supplementary feeding. Such feeds should contain at least 25% of animal protein, as the pond is likely to be unable to produce sufficient zooplankton to support a double stocking density. A built-in filter aerator unit has other advantages, mainly in extreme climatic conditions – *eg* when water temperatures are higher than 25°C or where very low temperatures in winter result in ponds freezing over for long periods.

10 Intensive carp farming in warm water

One of the latest developments is to grow carp in warm waste water from power stations. Power stations produce huge amounts of warm water, which cannot be recycled and is therefore wasted. In recent years the Central Electricity Generating Board has invited various industries to make use of this water, and at the present time it is being used for a variety of purposes *eg* vegetable growing, eel farming and carp farming. Such developments are also taking place in other parts of the world.

Carp can be grown at a very fast rate in temperatures of 23–25°C, but before any attempt is made to seek permission for carp growing in this way, a number of points must be considered:

- is the water from the power station in question suitable? (water extracted from industrially polluted rivers and brackish water is not suitable).
- are trained staff available? (the manager should be qualified to degree standard in biology or chemistry).
- has the market been secured? (in water from power stations table carp will be produced all the year round. European people eat carp only around Christmas time. It is the Chinese market which is interested in an all year round supply. Carp grown in water from power stations are unsuitable for angling purposes unless they are very gradually brought

back to natural conditions).

- can the unit be made large enough to make it viable? (a 100 tonne per year unit should be considered a minimum).
- is the supply of young fish guaranteed? (carp used for rearing in power station water should be produced in a laboratory; supplies from open ponds can introduce parasites and diseases, which develop very rapidly because of the high stocking densities

Fig 64 Healthy C1 fish suitable for growing on in heated water

Fig 65 Substandard C1 not suitable for growing on in heated water

involved. Conditions are somewhat harsh in the growing tanks, therefore only fish in good condition should be used, deformed fish are not suitable).

The function of such a unit is as follows: water is pumped from the cooling tank into a mixer tank where automatically the water is mixed to the desired temperature. The automatic regulator holds the temperature at a reasonably steady level, ideally 23–25°C, the water is then distributed to all tanks. When calculating stocking density, the size of the tank and the rate of the water flow must be considered, maximum stocking density is 100 kg of fish to one cubic metre of water. The water flow for these figures should be at a rate of a complete water change every 6 minutes, but the flow can be lowered if the stocking density is reduced.

Water enters the tank on one side and flows out through a grid in the centre. Standard circular trout rearing tanks are the best to use; extra oxygèn is supplied by a blower unit. The feeding of the carp depends on the size of the fish. Small carp should be fed every 30 minutes due to their rapid growth rate, but food must be available to all fish for 12 hours out of every 24. Food can be given by hand, but it is more

Fig 66 Growing tanks in a heated water unit – extra oxygen is supplied by a blower unit

Fig 67 An electronically operated feeder

efficient to use electronic feeding units, installed above the tanks. The size of the food pellet used should be increased to correspond with the growth rate of the fish, but care should be taken in stocking carp food as it has a shelf life of only three months. After this time its vitamin content deteriorates and it should not be used. The quantity of food to be given varies according to water quality and water temperature. As a guide – the amount of food to be given is approximately 2–2.5% of the body weight of the fish. In total 1.8–2.3 kg of food should produce 1 kg of carp flesh. Carp pellets and crumbs are manufactured in the UK; they are less expensive than trout pellets as they contain less animal protein.

A typical carp diet consists of:

oil	6%
protein	31%
fibre	6%
ash	10%
vitamin A vitamin D3 vitamin E	to be added

A number of trace elements are added to provide better growth and prevent deformities; it is advisable to discuss these with the food manufacturers. The

Fig 68 Layout of a heated water unit at a power station

Fig 69 Overflow pipe controls water level in the tank and can be lowered for rapid emptying

tanks should be covered with netting to prevent the carp from jumping out, and to deter predators; which apart from entering tanks and taking fish, also present a real threat as far as disease is concerned. If disease is introduced in this way, it spreads rapidly at such high stocking density. The overflow pipe, which controls the water level in the tank, works on a swivel action; in the event of disease treatment being needed, it can be used to lower the water level thereby reducing the quantity of chemicals necessary for treatment. The pipe is also used for lowering the water levels at harvest time and for cleaning purposes.

Carp farming in the warm water from power stations can be very rewarding, but as conditions in these circumstances vary so greatly, it would be dangerous to generalise here, and professional advice should be taken before attempting such a venture. If there is any difficulty in obtaining such advice the author suggests that the desired assistance will be forthcoming by contacting the Unit of Aquatic Pathobiology, University of Stirling, Stirling, Scotland; or the Unit of Aquatic Pathobiology, Aston University, Birmingham. Only if all the biological and technical factors are known can a unit be designed to work within the limits of a particular power station, otherwise conditions may prove totally unsuitable. In addition a prospective grower must seek the permission of the Central Electricity Generating Board and the Water Authority, and in some cases a planning consent of the local authority is needed, before a unit can be established.

It is not advisable to attempt to grow koi carp in power station water; as, although they are the same species they are less hardy than common carp. This is probably due to inbreeding over many generations to achieve better coloration. Furthermore there is total absence of zooplankton in power station water, and the pigment in zooplankton improves the colour of the koi considerably.

There are a number of other industrial installations which may have a surplus of waste warm water, sometimes of a very high quality, but unfortunately such water is usually limited in quantity and insufficient for a viable carp farming venture. Recycling this water through a filter unit could be considered, and is usually carried out in two stages. Firstly the rough waste is extracted by leaving the water to stand in a large settlement tank. In the second stage the water is passed through a biological filter, usually gravel or a special plastic material manufactured in such a way that it has the largest possible surface area for bacteria to colonise. The bacteria feed on the waste and the purified water is returned to the tanks. A small amount of fresh water (approx. 10%) should be added throughout this process. The biological filter unit should be roughly half the size of the fish tank, there are a number of European manufacturers who can supply such units and these are available in various sizes.

11 Diseases and parasites

All animals are susceptible to diseases and fish are no exception. In many cases losses occur due to ignorance on the part of the handler, and breeders, growers, importers, hobbyists and anglers are all guilty to some extent. It is impossible to describe in this book all known fish diseases. Some can only be identified by growing them as cultures in a laboratory (*eg* virus diseases). In England we are fortunate that the Ministry of Agriculture, Fisheries and Food operate one of the best equipped laboratories in the world, and the Ministry makes every possible effort to eradicate all serious diseases. For this reason it is an offence to import fish without a licence. It is almost impossible to obtain a licence for the import of fish, and it is anticipated that regulations concerning the import of ornamental fish, will be more severely controlled in the future. Only ornamental fish can be moved freely within the UK. To move indigenous fish, permission from the water authority of the receiving area must first be obtained.

However, it can be said that outbreaks of serious virus diseases in Britain are very rare, and carp in particular is a fish which is not generally considered to be disease prone. Parasitic attacks and bacteriological diseases can be treated, providing they are accurately diagnosed. One of the golden rules here, as with the keeping of all livestock is that prevention is better than cure. A well cared for and well fed fish will be more resistant to disease and the bacteria present in the pond will not adversely affect healthy fish. Overstocking will increase the possibility of infection as stress which results from the crowded conditions renders the fish less resistant to infection.

The following rules should be strictly observed if disease is to be avoided:

- When liming the ponds, care should be taken that the correct amount of lime is used. Insufficient lime could result in some of the bacteria and parasite eggs not being destroyed.
- When manuring the pond, only well rotted manure should be used to avoid build-up of ammonia.
- Before filling the pond, a water test must be taken to ensure that no toxic matter has entered the supply (land drainage water may carry toxins from spraying).
- The pond should not be stocked before zooplankton have fully developed.
- After the pond has been stocked with fish regular inspections should be made (three or four times a week).

The presence of a dead fish in the pond from time to time can be considered normal, especially in fry and C1 ponds. Multiple deaths are however a much more serious matter, and in such cases a water test must be carried out as a matter of urgency. If the cause is found in the water (a lack of oxygen, high ammonia and nitrate levels), all feeding should cease immediately and the water oxygenated, either by circulating (if possible make a fountain) or by using a compressor or bottled oxygen to aerate the water via a diffuser.

If tests show that water quality is satisfactory then it can be assumed that disease organisms are the cause of fish mortalities. Any of the following behaviour is a sign that something is amiss:

- fish come to the surface and towards the banks, appear listless and make no attempt to swim away when approached;
- fish lying on their side or observed rubbing against foliage or roots;
- fish changing colour (becoming darker);
- fish not feeding.

Individual fish must be carefully examined for the presence of fungus around the mouth or on the fins; film over the eyes; white spots on fins, tails or all over the body; spongy growth on mouth or fins. If a microscope is available this can be used to examine a single piece of gill plate for parasites. All the previously described symptoms are an indication of the commonest carp diseases and, if correctly diagnosed, are treatable.

In more serious conditions where fish are found dead and dying in large numbers, bloated and with the scales protruding from the body, with ulcerations on the body or around the mouth, or with a gill plate missing; viral infections are the likely cause. Many of these are notifiable diseases and it is advisable in such cases to contact the water authority who will then take the necessary action.

However serious fish diseases may appear to be, they cannot be transferred to man.

Ichthyobodo necator or *Costia necatrix*

This parasite is so minute that it cannot be seen with the naked eye. It attacks fry and young fish, but can be carried by large fish without any apparent effect on them; this is possibly the reason that it is so easily transferred. It has been known to suddenly affect fry in a laboratory with no fish other than the brood stock present. Under the microscope the parasite has the shape of a kidney, and has small arms with which it attaches itself, usually to the gills. Fry and young fish which have suffered an attack darken and the gills lose some of their colour and become milky. Treatment is relatively simple; the fish should be lowered into a salt bath (2% salt), using a fine net, and submerged for 10 minutes. During this time the water should be well aerated. After the treatment the fish should not be returned to the original tank. A formalin bath can also be given to fish with this condition, and is prepared using 30 ml of formalin to 100 *l* water, the formalin being diluted to 35%. When using this method the fish should be submerged for 30 minutes.

To treat affected fish in the pond is difficult as the quantity of chemicals required cannot be estimated accurately, due to variations in pond conditions, and the presence of algae and plant life. The most effective method is to remove the fish from the pond, and as described before, transferring them to another pond after treatment. The 'infected' pond should then be limed to 8.5 pH, and left to stand for 1–2 weeks before restocking with fish.

If for any reason it is necessary to treat fish in a pond, a solution of malachite green can be used (3 mg malachite green to 4.5 *l* (1 gal) of water.) The solution should be evenly poured into the pond which should be aerated during the treatment and after 10 hours fresh water should be added slowly to dilute the solution.

White spot (Ichthyophthirius multifiliis)

This disease is prevalent in tropical fish and is dreaded by every aquarist. It is recognised by the appearance of white spots on gills, fins and tail and in severe cases the whole body of the fish can be affected. The spot can grow to the size of a pin head, but consists of only one cell. The organism grows to maturity in the layers of fish skin, and then flakes off and falls to the bottom of the tank or pond, where it divides to produce 500 to 2,000 young.

These tiny protozoa lift off and float in the water, and if they attach themselves to a fish, another parasite develops and the chain is repeated. The nature of this cycle ensures that the disease develops much faster in

Fig 70 White spot

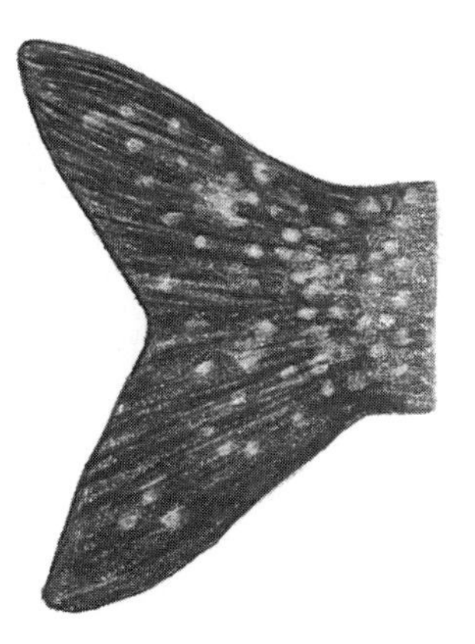

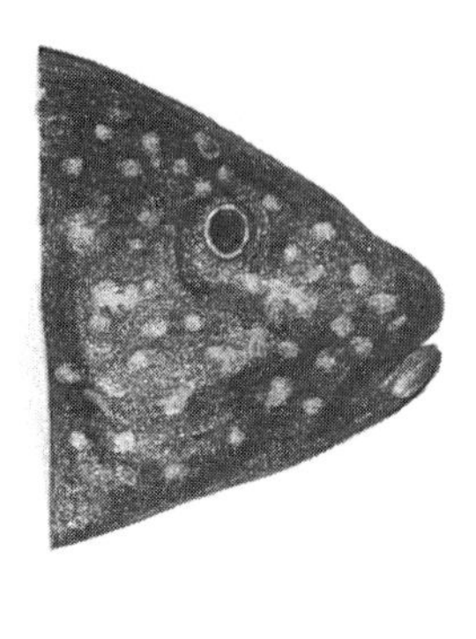

tanks than in ponds. The first sign of this disease is when fish can be seen rubbing themselves against plants and roots, as a result of the spots causing discomfort to the fish. As it is not possible to kill the parasite whilst it is attached to the fish, it has to be treated in its free-floating stage. A flow of fresh water through the pond will eradicate the protozoa or alternatively a malachite green treatment (a solution of 2 g/10 m^3) should prove effective. The malachite green treatment should be repeated after 1 week using 1 g/10 m^3 water. When treating in tanks, it is recommended that commercially prepared white spot treatment solutions (obtainable from most pet shops) be used. If not available, a solution of 0.2 ppm malachite green solution can be given. It should be remembered that *zinc free* malachite green should be used in all treatments.

Eye fluke (Diplostomum spathaceum) The eye fluke is a trematode which can affect carp. The parasite has a very complex life cycle in which it passes from snail to bird, from bird to fish and from fish back to snail. In the fish this parasite affects the eye (hence its name), and results in blindness, which of course, affects the fish's ability to find food, resulting in eventual starvation and death. Eye fluke is most common in ponds which have a large residential population of birds, particularly wild ducks and seagulls. If it is decided to keep ducks on a pond, they should be of the domestic type, and not clipped mallards, which can introduce the disease. Seagulls also can be controlled (see *Predators*).

Gill fluke (Dactylogyrus vastator)

Dactylogyrus vastator belongs to a group of parasites which are particularly nasty. This parasite attacks the gills of the fish and unfortunately is present in almost all open waters, it is not usually harmful to larger fish but can do tremendous damage to young fish, and if not detected early enough, can kill every fish in a pond. When mature, *D. vastator* is approximately 1 mm in size, and if an attack is in progress it can be seen with the naked eye. The worm like creature sucks itself on to the gill plate and then drives little hooks into the gill for firm attachment, after about a week of feeding on the gill plate it becomes mature and begins to lay eggs, which fall to the bottom, where after 3 days, depending on the water temperature, the larvae begin to hatch. These then propel themselves through the water and eventually become attached to a fish. Failing this it can survive for about 4 days in normal water temperatures, but for considerably longer when temperatures are lower. If the larvae attaches to a fish, its life span can as much as double.

It can be seen that this disease is extremely difficult to control. The long free-swimming stage of the larvae ensures that the gill fluke is easily spread. Even a single drop of water transferred from one pond to another by water fowl or amphibians, will be sufficient to spread the disease organism. Not surprisingly a fluke infestation can spread readily with the transfer of fish, particularly when brood fish are removed to Dubish ponds or to the hatchery for induced breeding. As a preventative measure, all brood fish should be treated for this disease.

The most vulnerable time for an outbreak of gill fluke is from late June to mid-July. Fry and fish over

75 mm (3 in) in size are not affected and fish in well managed ponds, rich in zooplankton, can survive an infestation without losses, while the fish farmer may not even be aware of the situation.

Treatment of fish in the pond is always a last resort for a number of reasons:

- it is difficult to assess the amount of water in the pond and hence the treatment dose required;
- the presence of plants and algae can decrease the effectiveness of the treatment;
- the mature parasites and its larvae are relatively easy to destroy but not the eggs.

To achieve a successful treatment the pond must be dosed a number of times but this has an adverse effect on the zooplankton population, which may collapse after repeated treatments, depriving the fish of their natural food. The carp do not then take readily to dry food, which if left uneaten accumulates in the pond resulting in further problems (*eg* deoxygenation).

An attempt to treat gill fluke in the pond should be carried out as follows:

- Dissolve 0.3 mg of Dipterex in 1 m^3 of water in a suitable container and apply by evenly spraying it over the water surface, using a boat if necessary.
- Repeat this treatment after the fifth day.
- On the seventh day apply a 50% malachite green solution (0.2 ml/m^3) again sprayed evenly over the pond.
- Repeat this treatment on the eleventh day.

After the treatment careful feeding with trout crumbs of the appropriate size is recommended. When treating a pond without fish, the pond should be drained, limed and left to stand for 2 weeks before refilling with water.

A bath treatment using 40% formaldehyde solution (250 mg/*l* water) with a 30 minute period of immersion (or shorter if the fish show signs of distress), can also be employed after treatment, if fish are placed in a pond with a good supply of zooplankton, they will grow normally.

All brood fish should be given the above treatment before being used for breeding purposes.

An alternative bath treatment uses simple cooking salt (2.5 kg salt/50 *l* of water) immersion time – 5 minutes. In all treatments of this kind care should be taken to ensure that the fish are immersed in the solution for the recommended time, and then dipped for a few seconds in fresh water, before they are released. Bath containers should be well aerated.

Gyrodactylus An attack of this parasite is almost as bad as that from *Dactylogyrus*; except that instead of attacking the gills it attaches to the skin of the fish, and clings to it by means of small hooks driven into the flesh. *Gyrodactylus* produces live born young instead of laying eggs.

The parasite feeds on the skin of the fish, which becomes inflamed and patchy and in severe cases, slimy looking. Fins and tail can suffer attacks which are so severe that only the bony parts remain. Treatment for this condition is the same as that employed for *Dactylogyrus vastator*.

Fish pox The name of this disease is misleading, having nothing whatever to do with pox. For many years experts were baffled by this disease, but now it has been established that it is a viral tumour, and being similar in its development to the *Herpes* virus it is genetically as well as infectiously transmitted. Fortunately it is not a virulent virus and can perhaps be compared to cold sores in humans, which usually make a sudden appearance after a cold, and then disappear again. Similarly in fish it usually breaks out as a result of adverse conditions in the pond. The sores appear on mouth, fins and tails as gelatinous blisters. The condition often disappears again if good fresh water is kept running through the pond or, better still, if fish

Fig 71 Fish pox

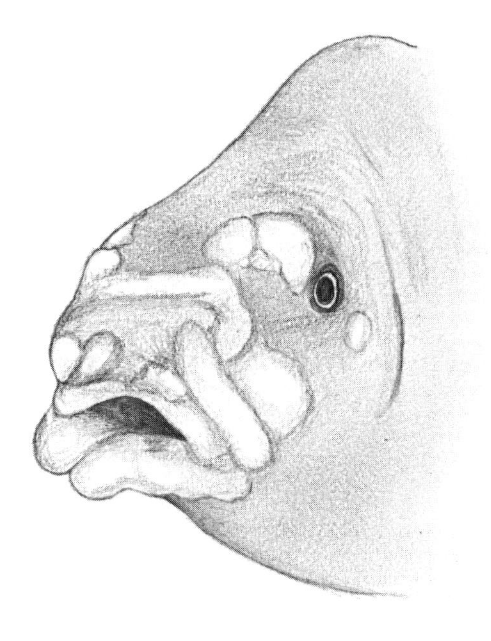

are transferred to another pond. This infection does not appear to have much permanent effect on the fish, they continue feeding and the infected skin areas recover quite rapidly. Where ponds have been emptied because of an outbreak of pox, they can be used again after lime treatment.

Other parasites

Ergasilus Fortunately this parasite does not attack carp and it is only mentioned here in the event of other species of fish being present in the carp pond, such as tench (see section on tench). *Ergasilus* is closely related to the cyclops family, and in fact has the appearance of a miniature cyclops with an egg sac on each side. It has vicious pincers on the head which it uses to attach itself to the gills. The best treatment for this condition is a 15 minute bath in a salt solution (20 g/*l* of water).

Fish leeches Usually introduced by water fowl this parasite prefers neglected ponds as it can survive in the mud for prolonged periods without feeding. When it comes into contact with a fish it attaches itself by means of suckers from both ends of its body. It can grow up to 4 cm and its movement is caterpillar like; reproduction takes place by means of eggs which are layed in cocoons. In low numbers these parasites cause relatively little damage to the fish and in summer the fish attempt to remove them by rubbing against stones or other solid objects. Continental farmers place sticks in the water to give the fish some chance of freeing themselves of this pest. This parasite is not particularly noticeable in the pond normally, but when emptied, their presence is very obvious by the way that they suddenly start to cling to the fish as the water is removed. The greatest threat to the fish from this parasite is not so much the injury it causes but by moving from one fish to another to suck blood, there is a great risk that any virus diseases present in the pond will spread through the fish population. It is claimed that tench will feed on leeches therefore a number of these fish in the carp pond could be beneficial.

When leeches are present they should be removed by a lime bath before the carp are transferred to wintering ponds and if necessary again after wintering, before removal to the growing ponds. To carry out this treatment two medium sized tanks are required – one to hold plain fresh water, the other the lime solution. Builders' lime is the most suitable and should be prepared in a bucket using 2 g lime to 1 *l* of water the night before treatment takes place. The lime must be well mixed until all the lumps dissolve. The solution is poured into the treatment tank, about 100 *l* is required in all, therefore 200 g of lime must be prediluted. The fish should be dipped for 5 seconds into the solution, lifted out and shaken in the fresh water for a few seconds before being returned to the pond or other container. Some leeches will not fall off immediately, but all will die after some time. Care should be taken not to exceed the 5 seconds of treatment, as a longer period harms the fish, particularly their eyes. It is not advisable to use the lime solution for a long time as it will quickly become diluted.

If during the growing period a heavy infestation of leeches is apparent, the pond should be sprayed with a lime solution (120 g of lime per hectare or 50 g per acre).

Fish louse The fish louse is not very common in the UK, but is often found on fish farms in Europe. It belongs to the crustacean family, and grows to the size of a lentil. The louse is almost transparent and is usually found on the fins and tail of the fish. It lives by sucking blood and is therefore strongly suspected of spreading virus diseases in the pond. It is particularly dangerous to young fish and 10–20 lice are capable of killing a fish of C1 size. Reproduction is by egg laying, after hatching the parasite must pass through various larval stages before reaching maturity. The louse is rather difficult to eradicate; 10 minutes in a bath of $KMnO_4$ (Potassium permanganate) at a concentration of 1 g/10 *l* of water is recommended, using a similar procedure to the treatment for leeches. Thorough liming of the dry pond will destroy both eggs and adult parasites.

12 Predators

Carp fry and young fish have many predators, mainly insects or their larvae. The following are among the most common:

- dragonfly larvae;
- damselfly larvae;
- backswimmer (*Notonecta glauca*);
- water scorpion (*Nepa cinerea*).

Unfortunately the carp farmer will always suffer losses from these predators but these can be minimised by not filling the pond with water too early before stocking; and after filling by introducing a plentiful supply of zooplankton from a special zooplankton

Fig 72 Great diving beetle (*Dytiscus marginalis*)

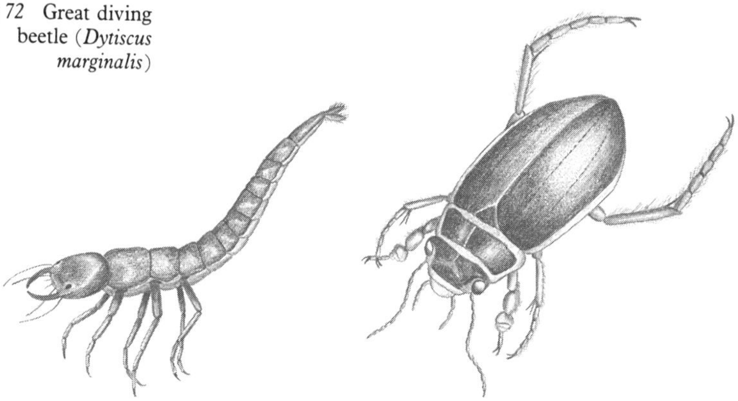

larva – up to 50 mm long

adult male – 30 mm long

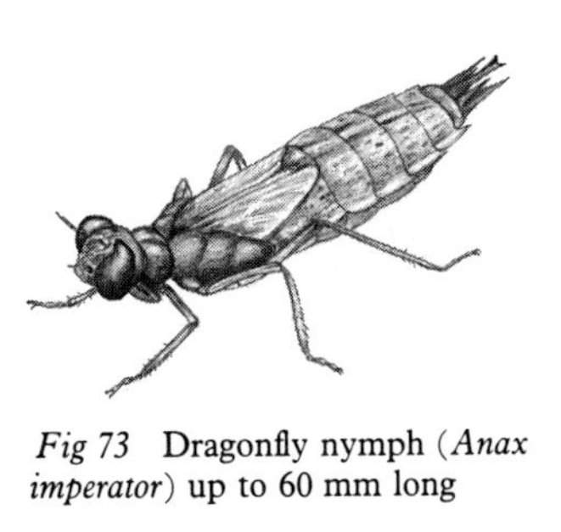

Fig 73 Dragonfly nymph (*Anax imperator*) up to 60 mm long

Fig 74 Damselfly nymph (*Enallagma cyathigerum*) 30 mm long

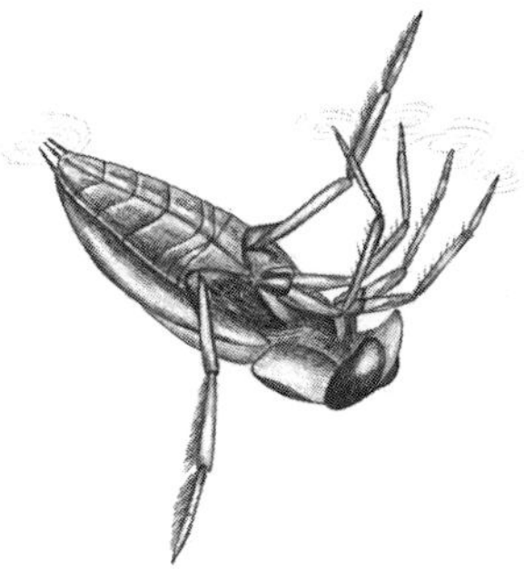

Fig 75 Backswimmer (*Notonecta glauca*) 16 mm long

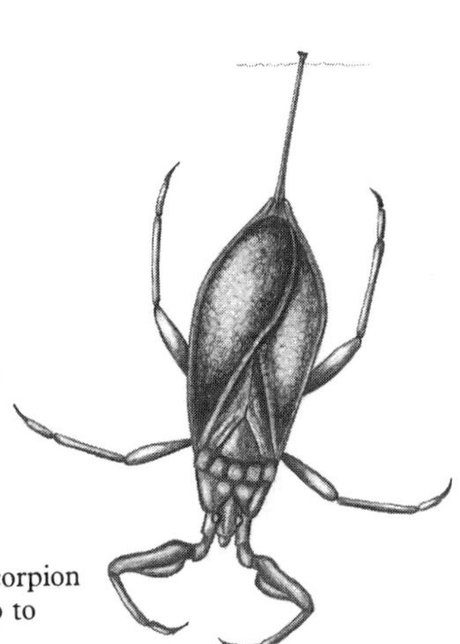

Fig 76 Water scorpion (*Nepa cinerea*) up to 35 mm long

pond. This will considerably shorten the time period available for the pond to develop, and consequently there is less time for predators to establish themselves.

When the fish have grown to approximately 5 cm (2 in) in size they are safe from these predators, but others are waiting.

Kingfisher This beautiful bird can do considerable damage in fish ponds. But as it is a protected species, the only way of dealing with the problem is to make the taking of young fish as difficult as possible for the bird. The use of humline will give some protection; this is a tough polypropylene tape which, when stretched lightly between stakes or canes, gives out certain vibrations that the kingfisher, and other birds, cannot tolerate. The only disadvantage to this method is that, when there is no wind, the tapes do not vibrate but as they are not expensive, this method is worth trying.

Fortunately for the carp farmer, the young fish will quickly become too plump for the kingfisher to swallow easily, and in fairness to the bird, it has to be acknowledged, that it also feeds on sticklebacks, water beetles, dragonfly larvae and a number of other large insects – all of which are unwelcome in the carp pond.

Sea gulls The blackheaded gull can become a real nuisance, and may even develop diving routines in order to catch a fish. There are battery operated systems available to scare these birds, similar to those used by airport authorities, and these work very well; they are however very expensive. It has been found that, a much simpler and less expensive way of dealing with this particular problem is, to stretch nylon string zig-zag fashion from pegs 1 m (3 ft) apart, over the pond. This irritates the gull and is a very effective deterrent (see *Fig* 77).

Crested grebe This very handsome bird feeds mainly on fish, but also takes frogs, newts and water beetles. It is a very shy bird and will not tolerate the regular presence of man, therefore it is much more at home on very large lakes.

Dabchick The dabchick is one of the smaller water fowl and belongs to the grebe family. From a distance it is relatively easy to mistake it for a moorhen, but it is in fact slightly smaller. It is an excellent diver and has very good eyesight. Because of this, it is very difficult to detect as it spots man from a long distance and then dives. Appearing again for a short breath, before

Fig 77 Humline on bamboo canes; nylon lines on short pegs criss-cross the pond

disappearing once more. The bird feeds on small fish, large water insects, as well as some vegetable matter.

Wild duck Most wild ducks which visit freshwater ponds are mainly vegetarian, but ducklings in a fry pond are disastrous. As they are very fast growing, needing large quantities of protein in the early stages of life, they take large quantities of insects and fry.

Heron The heron can be regarded as the master predator and it is both feared and hated by trout farmers and carp farmers alike. However it is deserving of some respect for its outstanding capabilities; its extraordinary eyesight renders it capable of fishing from dawn to dusk and even by full moon at night. Even at great heights this bird can see the slightest movement in the water, it will patiently stand for hours on one leg observing all movements in the area. Also it will soon reach the conclusion that a scarecrow presents no threat nor does the man driving the tractor in a nearby field, provided the tractor keeps moving. The heron will even take all the goldfish from a garden pond near to a house in the early hours of the morning.

Although fish, and especially eels, are the heron's

Fig 78 Criss-cross nylon strings are very effective against herons

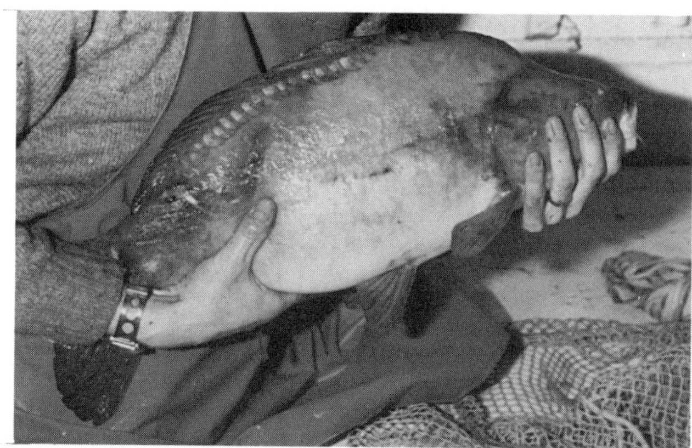

Fig 79 The scar on the back of this breeding female was inflicted by a heron

Fig 80 Killed by a heron – gulls and crows do the rest

main diet, it will take a variety of other prey, such as mice, voles, rats, moles, young rabbits and even moorhens, coots, snipe and ducklings are not safe from this bird. The heron rears 3–5 young, therefore large quantities of food will be needed during the rearing season, and the hunting periods are prolonged. The heron is a protected bird and shooting is only permitted in cases where it can be proved that they are a real pest. In the author's view, shooting serves little purpose as the heron is a territorial bird and when a territory becomes vacant, another bird will appear to take its place. The most difficult ponds to protect against the heron are shallow ones and those with shallow banks. If a pond is correctly constructed, the banks will have an angle of 45 degrees making it very difficult for a heron to fish. Humlines and criss-cross nylon strings, as previously described for the kingfisher give some protection and small ponds can be adequately protected by netting over.

Cormorant

The cormorant is basically a sea bird but can be seen occasionally as far as 30–40 miles upriver from the sea. It is unfortunate if this bird discovers fish ponds near to large rivers, as it is capable of swallowing fish of an astonishing size. Fortunately its appearance inland is rare.

Otters and wild mink

The otter is now so rarely seen that it is reasonably safe to assume that the carp farmer will not be bothered by it, but if it should appear, the nearest conservation group should be contacted, and they will arrange for the otter to be trapped and relocated.

There have been reports in recent years that wild mink are on the increase. They are excellent swimmers and so present a danger to fish farmers. The reason for their increase is in some part due to the action of organised groups releasing these creatures from mink farms, and the mink becoming wild as a result. Again, the way to deal with mink is by trapping.

In conclusion the would be carp farmer should not be deterred by this section on diseases, enemies and predators, which may have been written in such detail as to dissuade the potential aquaculturalist. This is not the intention but the author feels that as much information as possible on this subject should be given; with good husbandry and management the information will only be beneficial. Hopefully such problems will not arise, but if they do, the reader should be confident that they can be dealt with.

13 Other fish in the carp pond

Carp tolerate a large number of other species in the pond, but whether it is wise to mix them together is debatable. Carp outgrows all the other pond fish and this presents a problem at harvest time or when netting, as in the confines of the mesh the plump and heavy carp may cause injury to the smaller species.

Tench (*Tinca tinca*) In limited numbers tench are regarded as a welcome addition to the carp pond. Even more of a bottom dwelling fish than the carp, they feed on almost everything they can find in the mud of the pond, it is said that tench will even consume carp faeces. As the fish needs less oxygen than carp, about 10% of tench can be added to the pond without upsetting the biological balance. As a table fish the tench has little value, it grows at half the rate of the carp.

The tench is a popular fish with anglers and there is a growing demand for the fish in the ornamental market – they are said to help in cleaning the bottom of the garden pond. In instances where carp ponds are repeatedly troubled with leeches, tench are an asset, as they will feed on these parasites.

Tench are easy to sex as the pelvic fins of the male are almost twice the size of the female (see *Fig 81*). The fish tend to seek seclusion and therefore prefer to spawn in small, well overgrown ponds. Breeding takes place in late May or early June depending on the

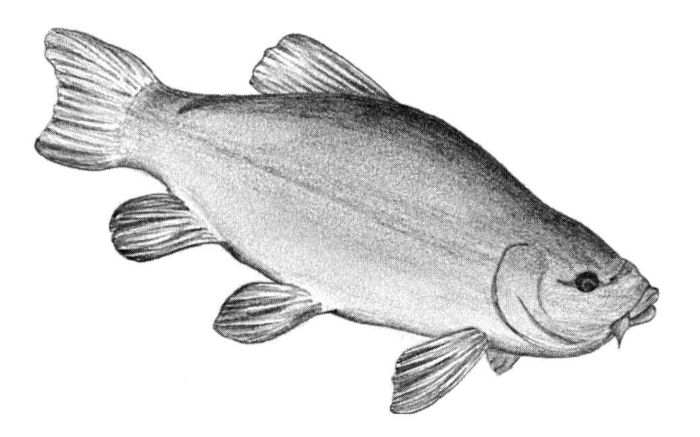

Fig 81 Tench (*Tinca tinca*)

temperature. Induced breeding is the same as for carp, and carp pituitary, which is more readily available, can be used in the breeding of tench. A large female carries up to 3,000 eggs.

Grass carp (*Ctenopharyngodon idella*)

The grass carp came to Europe as recently as the 1950s, the fish originates from the vast waters of the Amur in China, hence it's name – the White Amur. As this fish is a vegetarian, it can be invaluable in cleaning up overgrown ponds, dykes and lakes, and it was for this reason that the UK Ministry of Agriculture, Fisheries and Food (MAFF) decided to experiment with the grass carp, and at the time of writing, it is believed that these experiments are continuing. The main difficulty with this fish is its requirement for warmer conditions than *Cyprinus carpio*. It effectively begins feeding at 15°C and at suitably high water temperatures will consume vast quantities of water plants. From 50 kg of plant material, the grass carp will only put on 1 kg of body weight; this alone indicates the necessity for this fish to feed continuously. It will not breed in ponds in the UK, but has been successfully reared using induced breeding methods. Any movement of grass carp in the UK requires the permission of the Ministry of Agriculture, Fisheries and Food.

Trout Of the salmonids only the rainbow trout (*Salmo gairdneri* R) can be stocked together with carp, and then only where the carp pond is more than 2 m (6½ ft) deep at one end. The rainbow trout is in many ways more tolerant than the brown trout (*Salmo trutta*). The former has a lower oxygen demand, can withstand higher water temperatures and is tolerant of a certain amount of cloudiness in the water.

Stocking rainbow trout and carp together is sometimes possible in gravel and sand pits particularly if one end of the pit is shallow and the other very deep. The carp will occupy the shallow end which will be warmer in summer and hence there will be a plentiful supply of zooplankton and plant life. The rainbow trout will tend to remain at the deep end where the bottom temperature should not exceed 20°C – the highest temperature rainbow trout can tolerate. Stocking density will vary and is largely a matter of experience but if carp is to be the main stocking fish, not more than 120–250 rainbow trout per hectare (50–100 per acre) is recommended.

Signal crayfish (*Pacifastacus leniusculus*) In recent years much has been said and written on crayfish farming, it is therefore, not surprising that the author has been asked whether or not crayfish can be stocked with carp. The basic idea is an attractive one, crayfish are a delicacy and command high market prices. Furthermore the two species live compatibly together. The difficulty is that in order to keep carp ponds in maximum production they need to be emptied, limed and fertilised each year, and because crayfish tend to live in burrows they could not all be recovered and a large number would perish at these times.

There are however, some ponds which cannot be emptied yet still bring a reasonable return when stocked with carp. Such ponds would be suitable and crayfish could be stocked in these circumstances.

The signal crayfish is a native of California. It was

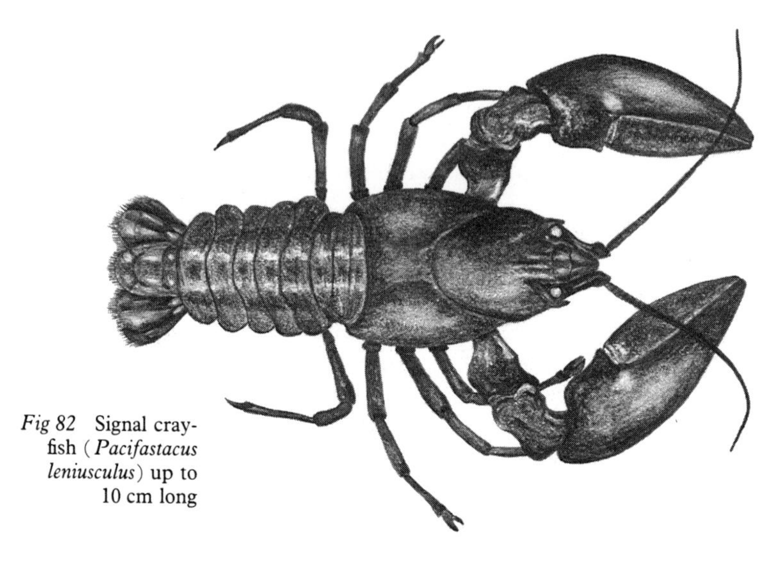

Fig 82 Signal crayfish (*Pacifastacus leniusculus*) up to 10 cm long

brought into Sweden in the 1960s after virtually the entire crayfish population of Europe perished as a result of the crayfish pest.

After a world-wide search for an alternative to the European crayfish, Swedish scientists discovered that the signal crayfish was immune to the pest. Experiments in Sweden were so successful that large numbers were imported. This was the beginning of a new crayfish industry and later, the signal crayfish arrived in England.

At the present time crayfish farming is expanding in the UK. The climatic tolerance of the crayfish is equal to that of the carp – it stops feeding during cold periods.

Young crayfish feed mainly on water plants but later in life will take earthworms, insects and insect larvae, snails and small fish – with a preference for sticklebacks. If food is scarce, they will feed on dead animal matter. The breeding process of crayfish is somewhat complicated, but they do breed in open ponds in the UK. In the last few years methods of stripping the fertilised eggs from the female have been developed, the eggs are then incubated in Zuger jars.

It is wise for the beginner to start by buying juveniles from a well known producer or importer, which can then be released into a pond. As the crayfish favours seclusion, it is a good idea to place a number of drainage pipes in the pond, which the crayfish will use for shelter. After three summers the crayfish is both fully grown and sexually mature and should be 10–12 cm (4–5 in) long. Trapping is the best method of harvesting, and as the crayfish is nocturnal, the traps should be baited with fresh pieces of fish and laid out in the evening, to be lifted the following morning. The harvested crayfish can be conveniently transported in plastic bags (see *Transporting carp*).

Finally it should be noted that as the crayfish grows it sheds its shell from time to time. At a moulting stage it is very vulnerable to predation, especially from large carnivorous fish. Not only is the body soft and easy to swallow, the crayfish is also very weak and not able to defend itself. Eel, pike, zander and perch will all attack the crayfish at this vulnerable time.

Eel (*Anguilla anguilla*) The eel begins its life in the Sargasso Sea off the east coast of Central America. It takes the small larvae about three years to migrate to European coastal waters. At this time the young eels are almost transparent and are known as elvers. Entering estuaries in large numbers they reach streams, dykes and lakes looking for suitable feeding grounds. Eels can even travel short distances through wet grass, and so can easily appear in carp ponds, particularly ones which are located near to streams and dykes. They are carnivorous and are capable of inflicting severe damage in fry and fingerling ponds. It is advisable to make regular checks for the presence of eels in these ponds by setting well baited traps or laying out night lines. Eels spend some 6–10 years in freshwater ponds, and the female can grow to a length of 1.30 m during that time. Eventually the urge to reproduce will force the eels downstream and back into the North Atlantic, which they have to cross once again to reach the

Fig 83 Eel (*Anguilla anguilla*) up to 1 m long

Sargasso in order to breed, after which, the eel dies.

Eels are regarded as a delicacy in Britain, particularly in London, which is famous for the jellied eel. In continental Europe smoked eel is regarded as a gourmet dish, and ½ kg (1 lb) of smoked silver eel (those which are returning to the sea) costs the equivalent of ½ kg (1 lb) of fresh smoked Scotch salmon.

As eels, like carp, grow more rapidly in warm water they are now intensively farmed in the warm waters from power stations, in the same way as carp. A number of organisations in Britain are already involved in the intensive farming of eels in this way, those which they produce are all exported to the continent of Europe, but it is not known how successful such ventures are.

Growing eels and carp in polyculture is not advisable as eels have a much slower growth rate than that of the carp, and the eel is carnivorous as previously mentioned.

Stickleback Which boy cannot boast of his success in catching sticklebacks, armed with perhaps no more than a large jam jar and a home made net made from a piece of wire and mother's old stockings! Sticklebacks are easy to find, they appear in almost any place where water is in abundance. They can adapt to life in fairly acid water, slightly alkaline water, into brackish and even sea water. When fully grown they are about 5–7 cm (2–2½ in) long.

Sticklebacks usually find their way into the carp pond by being carried in egg form on the plumage of water fowl. Where there is a plentiful food supply in the water, they reach maturity quite rapidly, and if a partner can be found, breeding begins without delay, but will only take place after a nest, using plants, roots or other suitable matter, has been built. The male, who does all the nest building, is by now in full breeding colour. This is very attractive; scarlet red on the underside of the body, a yellowish tinge on the head and dorsal fin, and blue colouring towards the tail. The male literally pushes the female into the nest, which soon fills with eggs, these have the appearance of mustard seeds. The nest is guarded by the male, who will defend it against anything which comes too close. The young sticklebacks start to feed on micro-organisms such as rotifers and protozoa for 3 or 4 days and after that small zooplankton is added to their diet. They grow rapidly and because of their very sharp spikes, which they erect when danger threatens, have very few enemies. This can result in sticklebacks overbreeding in the carp pond, which is detrimental to the carp which will eventually be deprived of essential animal protein. The biological balance in the pond can also be threatened if sticklebacks are allowed to breed unhindered, thousands may be produced in a single season, and in such numbers may seriously deplete the oxygen content of the water for all other species in the pond, particularly during a hot spell.

Thirdly, a very unpleasant side effect can occur in stickleback infested ponds when these have to be netted. The fish's spines are needle sharp and will

Fig 84 Nine-spined stickleback (*Pungitius pungitius*)

seriously damage other fish netted with them. It is therefore advisable, where sticklebacks appear to be present in excess, to take immediate action by trapping them out before any injury to other fish occurs. The trapping method to use is the same as that for small carp. The same traps can be used, providing that the carp in the pond are bigger than the entrance holes in the trap. The making of traps is described in the section on trapping and netting koi.

Bitterling (*Rhodeus amarus*)

From time to time the carp farmer will notice the presence of other species of small fish in the carp pond. These are mostly species native to Britain and Europe, and it is believed that water fowl are responsible for spreading the eggs, and even small fry, from pond to pond as they swim in the water. The swimming action washes the eggs and small fry from the bird's plumage into the pond. Usually these small guests present no problem apart from the fact that they are competing for the same food supply. They do not breed as vigorously as the stickleback and have no spikes with which to defend themselves, and therefore are more prone to predation; they include the following species: minnow, rudd, roach and bitterling. The latter is the most interesting – it is a very attractive fish which grows to only 5–7 cm (2–2½ in) these characteristics and the fish's placid behaviour make it as suitable for the aquarium as many tropical species – and indeed some enthusiasts keep bitterling in cold water aquariums, where they can even be bred. Somewhat carp-like in shape, the silver coloured body has a touch of shiny-blue shimmering through the scales, with the blue becoming more pronounced towards the tail. When in breeding condition, the male becomes strikingly coloured – a shiny steel blue to scarlet red – mostly on the lower part of the body and around the gills.

The breeding habits of this little fish are extraordinary; it needs a freshwater mussel (*Unio* or *Anodonta* species) to produce its offspring. The breeding season of the bitterling stretches from spring to early summer, and is started off by the male who first selects the mussel which he then carefully guards. In the meantime the female watches the procedure and as a result comes into breeding condition. She then develops a tube from her vent through which the eggs pass into the gills of the mussel. This tube or ovipositor grows to a length of 5 cm (2 in) and it is extraordinary how the female can find the very small breathing slit of the mussel with the ovipositor. What is more astonishing is, that the mussel does not reject the bitterling and accepts the 40–60 eggs laid into its gills. Furthermore it breathes in the sperm from the male bitterling to fertilise the eggs. The eggs are incubated for some 3 weeks inside the mussel, on hatching the fry remain for a further 3 days or so until the yolk sac is absorbed, after which time they leave the mussel to fend for themselves, feeding on small zoo- and phytoplankton. Carp farmers, who are fortunate enough to have bitterling in their ponds should make effort to encourage them to breed as there is a demand for these fish in aquarist shops. A story from the 1930s records that the demand in some countries for the little fish was so great, that high prices were paid for them; apparently they were used for pregnancy testing. It was said that if the urine of a pregnant woman was added to an aquarium containing bitterlings, the females would develop the ovipositor. The author cannot substantiate this story!

Summary Growing other fish together with carp in the same pond only makes sense when the other species present no danger to the carp and do not compete for the same basic food. Fish which feed below or above the carp's level, such as tench (feeds below) and orfe (feeds above) can safely be stocked. Orfe is a surface fish and can be grown with carp, but it has a distinct disadvantage in that it attracts too many predators, such as sea gulls and herons.

14 Harvesting

As winter approaches and the water temperature in the pond decreases, the carp stops feeding, and the carp farmer must prepare for harvesting. The emphasis here is on preparation and this should start with the checking of protective clothing, nets and holding tanks. The wise carp farmer will now be contacting his customers to discuss estimated delivery dates. If ponds are large, pre-netting is advisable; the advantage here is that not all the fish in the pond will need to be harvested at any one time, and some fish will already be sold before the pond is finally emptied. As the water temperature is already low, the carp is easy to

Fig 85 Netting

keep as oxygen requirements reduce with the fall in temperature hence the carp becomes lethargic, making it easier to handle.

Carp intended for the table market should be held for 2–3 days in storage tanks which have a flow of water running through, this cleans their intestines and gills. It is relatively easy to harvest the rest of the pond; the shutter sections to the monk should be removed one by one until all the water has been drained and the fish are in the catch-pit. The fish should then be removed by hand netting and transferred to oxygenated transport tanks. Harvesting must take place when the water temperature and the air temperature are more or less constant, fluctuations should not exceed 2°C either way. More drastic changes than this are stressful to the fish.

15 Transporting carp

The smaller the carp the less likely that problems will be encountered in transporting the live fish. The best method of transporting C1 size carp is by using plastic bags. These should contain water and oxygen and for extra safety, two bags should be used, one inside the other, with both finally placed in a suitable container. The standard size of these bags is approximately 90×45 cm; 6 *l* of fresh oxygenated water is placed in the inner bag, which should be aerated whilst the fish are being counted in. The next stage is to blow oxygen into the bags using an oxygen supply pipe which should then be pulled out, and the inner bag secured at the top. The way to do this is to tightly twist the top part of the bag about a quarter of the way down. The twisted part should then be bent downwards and secured by means of strong rubber bands. This procedure should be repeated for the other bag thus making a double seal. Experience will show the quantity of fish which can then be transported by this method; it depends on the size of the fish, the duration of the journey and the temperature of the water. As an example: to transport 2,000 2.5 cm (1 in) size carp by car for 5 hours in summer would be quite safe if several ice bags are packed around the carrying bags. With 7.5 cm (3 in) size, approximately 200 fish could be transported in the same way. When sending fish by rail, these numbers should be halved in case of delays,

feeding in oxygen

the bag is closed by twisting the end, bending it over and sealing with a rubber band

the sealed bag is packed in a strong cardboard or polystyrene box

Fig 86 Transport of small carp in plastic bags

which happen quite frequently. In these circumstances, fish may be left standing for quite lengthy periods on a railway station with no movement whatsoever, which means that the oxygen in the bag is not mixing with the water, and the carp are being deprived of the oxygen they need.

It is not advisable to transport large carp in plastic bags as the dorsal fin is very sharp and pointed and can easily pierce the bag.

When transporting fish, it is advisable to protect them from possible fin damage by adding a few drops of acriflavine to the water. This is a deep orange powder which is a well known antiseptic, can be easily dissolved in water and is effective on fish. Acriflavine powder can be bought without a prescription, and is prepared for use as a fish antiseptic as follows: mix one level teaspoon of acriflavine with 1 *l* of distilled water and shake until completely dissolved. Then add two drops of the acriflavine solution for each 4.5 *l* (1 gal) of water in the transporting bags. The strength of the solution can also be judged by the colour of the water, which should be light green. Acriflavine is harmless to the fish, and will produce no side effects even with a slightly higher dose than is recommended here.

Transporting carp of C2 and C3 size is usually carried out using large transport tanks. The standard transport container holds 1,000 *l* (approx. 250 gal). The container should first be half filled with water and a large diffuser connected to an oxygen cylinder with the supply turned on to give a steady flow of bubbles. The fish should then be loaded into the container before it is topped up with water to four fifths of its depth. The deeper the water in the tank the more diffuse will be the oxygen content. If prepared in this way, it is possible, in normal weather conditions, to transport 0.5 tonne of C3 size carp safely for 5 hours; but only about 400 kg (approx. 900 lb) C2 size. In winter these numbers can be increased by 20% but during warm spells and also in thundery weather, these figures should be halved. Acriflavine should be administered as previously described, but not to carp intended for immediate human consumption. Carp of C1 size can, of course, be transported in this way too, and the suggested stocking number, for the same normal weather conditions, is about 10,000 fish.

It is strongly recommended that a spare full cylinder of oxygen be carried on the transporter and also that a stop be made every hour to inspect the fish. Fish destined for a long journey should never be taken directly from the pond, but should be placed in a holding tank for at least one day beforehand. All feeding should cease to allow them to clear their intestines, otherwise they will foul the water in the transporter tank. A transporting vehicle should never be left to stand in strong sunshine whilst fish are in transit, as the temperature of the water will quickly increase, which in turn activates the fish considerably, and consequently increases their oxygen demand.

An oil free 12 volt compressor can be used instead of a cylinder supply. The compressor is connected to the battery of the transporting vehicle, and operated in the same way as bottled oxygen. Bottled oxygen is more effective than compressed air, so there should be an increased flow from the compressor. However, both, compressed air and bottled oxygen, when used in excess, are harmful to the fish and can result in inflammation of the gills. On long journeys occasional inspection should be made for signs of over-oxygenation. One such indication is that the gills of the fish appear a much deeper red than usual.

Transporting large fish to angling waters

This is best carried out in early spring when the water temperature is just around 10°C (50°F) or early autumn. It is not advisable to move carp into a strange pond too late in the year as they will not settle, and in familiarising themselves to their new surroundings, will use up vital energy which should be reserved for the approaching hibernation period.

16 Table carp

It is traditional for many continental housewives, particularly in eastern Europe, to serve carp at Christmas time in the same way as we, in the UK, serve turkey. With the passing of time, and the spreading of different cultures, not only from country to country, but from family to family, eating habits are showing natural change. The author was not surprised to observe carp being offered as a main course on the menu in a restaurant recently. What makes carp eating so interesting, is that the fish is suitable for use in the preparation of so many different dishes. For the housewife, tempted to try a carp meal, here are 20 European recipes from which to choose.

In addition to the following recipes, carp are also delicious when smoked. The carp should first be gutted and cleaned, then cut into steaks of about 250 g (8 oz) size and placed in a brine solution 280 g (10 oz) salt to 1 *l* (2 pts) of water to soak for 1 hour. The steaks should then be placed on hooks in a smoker whilst still wet, and cold smoked for about 5 hours at approximately 27°C (80°F) until the skin is dry and golden brown. The temperature is then increased to 82°C (180°F) and maintained for 1 hour.

Preparing a carp for the table

To prepare a carp for the table the scales must first be removed. The scales of the mirror carp are easier to remove as they are fewer in number. The fish should be gripped by the tail, preferably using a cloth for a better grip and then starting at the tail make an upward movement towards the head, using a blunt knife. The knife should be aimed at the underside of the scales and in this way the scales will be removed fairly easily.

Scaly carp can be handled in the same way. As a point of interest there is a special descaling tool manufactured for this purpose and is quite readily available in kitchenware shops all over the continent of Europe. It is so designed that it not only scales the fish but also collects the scales in a closed compartment.

In case of difficulty in obtaining this moderately priced gadget one such manufacturer is:

SCALEX,
Westmark Schulte and Co. KG,
D. 5974 Hercheid,
West Germany.

Using a sharp knife the carp should then be split open, working from the head to the vent. The cut should not be too deep as the carp has a large gall bladder which is located quite near to the head and if the gall bladder is pierced the liquid will escape and contaminate the flesh, giving a bitter taste. To gut the fish it should be held firmly down with one hand and the fingers of the other hand run into the body of the fish, round the rib cage carefully removing all the intestines. If it is preferred that the head should be removed first, carefully cut around the gills before severing the head from the body (using a chopper).

A sharp medium sized knife should be used for filleting. The headless carp should be laid on the table, the head end facing away from the filleter. Hold the fish firmly down and make an incision along the top side of the dorsal fin, right down to the tail. Follow this cut carefully through to the other side, avoiding cutting through the bones of the rib cage. A cut should then be made on the tail, alongside the anal fin to complete the separation. To remove the main bone turn the fish over and begin to cut from the opposite end along the spine and rib cage until the skeleton leaves the flesh.

The head and tail of the fish can be boiled to make fish stock for use in soups.

Carp recipes

Baked carp

4 cross-cut pieces of carp
4 oz sliced button mushrooms
1 tablespoon chopped parsley
half an onion, skinned and finely chopped
salt and pepper
¼ pt red wine
¼ pt water
1 level tablespoon cornflour

Method

Soak the fish in salted water, rinse and wipe well. Put in a greased overproof dish and add the mushrooms, chopped parsley, onion, salt, pepper, wine and water. Cover with a lid or foil and bake in the centre of the oven for about 30 minutes, or until tender.
Remove the fish, spoon the mushroom mixture over it and keep warm. Strain and retain ½ pt of the cooking liquid.
Blend the cornflour with a little cold water and stir in the cooking liquid. Put into a pan and bring the mixture to the boil, stirring all the time until it thickens. Cook for a further 1–2 minutes and adjust the seasoning if necessary.
Serve the fish coated with this sauce.

Fried carp

carp
slices of bread
a lemon
time: 20 minutes or more, according to size

Method

Clean and dry the fish.
Flour well and place in a pan and fry until light brown.

Lie the fish on a cloth to drain, and fry some three-cornered pieces of bread and the roes.
Serve the carp with the roes on each side of the dish.
Garnish with the fried bread and sliced lemon, and make anchovy sauce, with the juice of a lemon added, to eat with it.

Jellied paprika carp

(4 servings)
1½–2 lbs carp
1 lb onions
1 tablespoon paprika
pepper
salt
water to cover

Method

Wash the carp and cut into thin slices.
Peel and slice the onions very thinly and put into a pan with the paprika, pepper, salt, and enough water to cover. Simmer for 10 minutes.
Add the pieces of fish with a little more salt and continue cooking until just soft. Lift out the fish and arrange on a dish.
Rub the sauce through a sieve and pour over the fish.
Serve cold.

Stewed carp

1 carp
½ pt water
½ pt port wine
1 tablespoon of lemon pickle
1 tablespoon of browning
1 teaspoon of mushroom powder
1 onion
6 cloves
horseradish root
cayenne pepper
a large lump of butter
a little flour
juice of one lemon
time: 75 minutes

Method

Having scaled, cleaned and taken out the gills, wash the carp thoroughly by soaking it in water for 30 minutes and dry it in a cloth. Dredge a little flour over it, and fry until light brown.
Put it in a saucepan with the port wine, water, lemon pickle, browning, mushroom powder, horseradish root, a little cayenne pepper and the cloves stuck in the onion.
Cover the saucepan closely, so that the steam may not escape, and let it stew gently over a low heat until the gravy is reduced to just enough to cover the fish.
Remove the fish and place in a serving dish.
Reheat the gravy and thicken it with a lump of butter rolled in flour, then strain it over the fish and garnish with croutons.
Just before placing on the table, squeeze the lemon juice into the sauce.

Stewed carp

2 carp
½ pt water
½ pt white wine
a blade of mace
12 whole peppercorns
1 onion
a bunch of sweet herbs
a little salt
horseradish root
time: 90–95 minutes

Method

When the carp are well cleaned and washed, place them in a saucepan with the other ingredients. Cover the pan and simmer over low heat for an hour and a half.
In another pan place the following: a gill of white wine, two chopped anchovies, a little lemon peel, ¼ lb of butter rolled in flour, two tablespoons of cream and

a large teacup full of the liquid the carp was stewed in. Heat the sauce for a few minutes and drain the carp. Add to the sauce the yolks of two well-beaten eggs, mixed with a little cream; when the sauce boils, squeeze in the juice of half a lemon.
Serve the carp with the sauce poured over it. Garnish with crisp parsley and red pickles.

Blue carp (Carpe au bleu)

carp
vinegar
port wine
3 sliced onions
2 sliced carrots
parsley
2–3 bay leaves
3 cloves
thyme
pepper and salt
time: 1 hour or more

Method

Clean the carp well, but in doing so make as small an opening as possible.
Place the fish in a fish kettle of the right size. Pour over it half a pint of boiling vinegar, and add enough port wine to cover it. Add the remaining ingredients. Simmer gently over a low heat for about an hour. Remove from the heat, let the fish cool in the liquid and serve.

Blue carp (Karpfen blau)

1 large carp
1 sliced onion
a pinch of mixed herbs
peppercorns
wine vinegar

Method

Clean the carp on a wet surface and salt it on the inside only, so as not to disturb the slimy skin, which

gives the 'blue' colouring. Tie the tail and head of the fish together to make a circle.
Place the fish in a large pan and cover with vinegar. Bring to the boil and add the sliced onions, herbs and peppercorns. Simmer very gently until the fish is tender but firm.
Lift out and rinse well in hot water to remove some of the vinegar flavour, or lift out and douse in melted butter.
Put on to a hot dish and serve with creamed horseradish sauce or browned butter containing capers and parsley.

Carp in bier

(6 servings)
1 carp (approx. 3–4 lbs)
3 tablespoons white vinegar
2 chopped carrots
1 chopped leek
1 chopped onion
1 bay leaf
1 clove
salt and black pepper
6 oz gingerbread
1½ pts brown ale
juice and rind of lemon
time: 1½–2 hours

Method

Clean the carp and leave it to soak for 1 hour in cold water and vinegar. Put the chopped vegetables, bayleaf, clove, salt and pepper into a pan large enough to hold the carp. Add 1 pt of water, cover and simmer for 1 hour. Break the gingerbread into cubes and soak in ½ pt of brown ale.
Put the carp into the stock with the lemon juice and the remaining ale. Cook over low heat for 20 minutes, or until the carp is tender. Lift out the fish and keep it warm. Strain the liquid, add the gingerbread, and boil

this mixture rapidly until it has reduced by half.
Strain the sauce over the carp and garnish with lemon rind.

Stuffed carp

1 carp (approx. 4 lbs)
veal forcemeat
1 small chopped onion
1 bayleaf
2–3 slices of lemon
salt and pepper
2–3 tablespoons of red wine or port
1–2 oz blanched and slivered almonds (optional)

Method

Soak the fish in salted water, rinse and dry.
Stuff the body cavity with veal forcemeat. Sprinkle with the chopped onion, add the bayleaf, lemon, seasoning, about ¼ pt of water and the port or red wine.
Cover with a lid of aluminium foil and bake in the centre of the oven for about 45 minutes, or until the fish is tender.
Put the fish on a serving dish and keep it warm. Strain the cooking liquid into a pan and reduce slightly by rapid boiling. Serve the fish coated with the cooking liquid and sprinkled with almonds, lightly browned under the grill.

Carp with sour cream

1 carp
4 new potatoes
2 oz butter
½ pt sour cream
1 oz breadcrumbs
seasoning

Method

Wash and dry the fish, but remove the head.
Put a layer of thinly sliced potatoes at the bottom of the dish, season and cover with half the butter and half

the cream. Bake for about 20 minutes.
Place the fish on top and cover with the rest of the cream. Cook in the centre of a moderate oven.
Cover the fish with crumbs and continue to cook for another 10 minutes.

Carpe farcie

carp
a few mushrooms
parsley
shallots or an onion
the hard roe of the carp
2–3 yolks of hard boiled eggs
2 anchovies
salt and pepper
minced parsley
knob of butter
time: 20–30 minutes according to size

Method

Chop up the mushrooms, parsley, shallots or onion, and the hard roe of the carp (if it happens to have one). Half cook the mixture, and then mix it up with the yolks of hard boiled eggs, finely chopped anchovies, salt and pepper. If the roe is soft, mash it up and mix it in this forcemeat.
Scale, open and clean the carp; stuff it with the forcemeat, and sew it up. Roll the carp in oiled paper and grill it.
In a dish place some butter, blended with minced parsley, pepper and salt, and serve the fish upon it.

Carp a l'etuvee

carp
2–3 oz butter
a little flour
3–4 small onions
a bunch of thyme
parsley
1 bay leaf
1 clove of garlic
3–4 mushrooms

grated nutmeg
½ pt of broth or water
½ pt red wine
time: 60–75 minutes

Method

Scale and clean the carp, and cut into pieces.
Brown some butter with a little flour in a saucepan, and add the other ingredients. Stew the carp in this mixture until cooked. Serve it with toasted bread in the dish, and with the sauce from the saucepan strained and poured over it.

Carpe frite

a small soft-roed carp
lemon juice
flour
time: 12–15 minutes

Method

Choose a small soft-roed carp; open it down the back, press it open very flat and take out the roe.
Flour both the fish and the roe well, and fry in very hot fat until light brown.
Serve with lemon juice squeezed over.

Marinade de carpe

slices of carp
pepper
salt
nutmeg
mace
3 cloves
1 sliced onion
a sprig of sweet basil
juice of lemon
time: approx. 1 hour

Method

Cut fillets or slices of carp, and place them in a saucepan.
Steep the fish in a marinade made from the other

ingredients until it thoroughly imbibes the flavour of the seasoning.
Put it over the heat and when three quarters cooked, take out the fish, drain, flour, dip into butter and fry it.
Serve with crisp parsley; and with whatever sauce is preferred.

Carp sur le grill

time: 12–15 minutes, according to size

Method

When the carp is scaled and duly prepared, rub it over with oil and boil it.
Serve it on a bed of sorrel, or prepared vine leaves, or with a caper sauce; or it may be eaten with wine, oil or vinegar.

Carp a la maitre d'hotel

Cover the fish with finely chopped sweet herbs, oil, and lemon juice, then grill it.

Dressed roe of carp

3 or 4 carp roes
a tablespoon of vinegar
salt
powdered mace
pepper
white bread crumbs
rind of half a lemon

Method

Boil the roes in salt and vinegar, then mince them up and season with pepper, salt and a small quantity of powdered mace, a few breadcrumbs and the finely chopped lemon rind.
Bake in a casserole, place into scallop shells and serve.
time: 15–20 minutes

Paprika carp

2–3 lb fish
1 lb sliced onions
4 oz butter

¾ pt cream
2 teaspoons paprika
salt

Method

Fry the onions lightly in the butter. Place the carp on the onions and add the cream mixed with paprika, salt and pepper. Bake until tender.
Strain the sauce over the carp.

Boiled carp

time: 30 minutes

Method

Scale and remove the gills from the carp, and rub some salt down the backbone; then lay it for half an hour in strongly salted water, which will thoroughly cleanse it. Dry and place in a fish kettle of boiling water, with a tablespoon of salt. Boil for 30 minutes (or less for a small fish).
Boil the roe with it, and serve garnished with parsley and slices of lemon.
Serve with plain melted butter and fish sauces.

Nuremburg carp

carp
salt and pepper
plain flour
horseradish or tartare sauce
time: about 15 minutes

Method

Clean, fillet and dry the fish.
Season, roll in plain flour.
Deep fry in very hot fat until brown and crispy.
Serve with horseradish or tartare sauce.

17 Summary of the carp farmer's year (in north European climatic conditions)

January Pond maintenance to be carried out. Check that rain or seepage water is flowing freely through the monk, ensure that water does not collect in the wintering ponds.

Check water in wintering ponds which contain fish, to ensure there is sufficient oxygen. If the levels are low, the freshwater supply should be increased.

If the ponds are completely frozen, ventilation holes should be cut out. Snow and slush should be removed from the ice if possible.

February As for January. The C1 supply for introduction in spring should now be ordered (if this has not already been done). Towards the end of the month, the ponds should be manured.

March Ponds should be filled with water and zooplankton introduced, if possible. Stocking may begin at the end of this month if the long term weather forecast is favourable.

April All ponds may be stocked where zooplankton have fully developed.

May Supplementary feeding can be started if the temperature is high enough. Check the fish for parasitic infection.

June Supplementary feeding should be increased, and if zooplankton populations show signs of decreasing, boost production with the application of super phosphate (250 kg per hectare or 100 kg per acre). Feeding must stop if temperature reaches 25°C.

July As for June. Care should be taken to ensure that required oxygen levels are maintained. pH and ammonia levels to be checked twice weekly, if possible, in the morning. Do not overfeed.

August In spells of hot weather (especially in the mornings), check for signs of distress in the fish – usually the result of an oxygen shortage.

September Feeding should be reduced as temperatures fall, C3 size carp should now be of table size. If not, a protein supplement should be given to boost growth. Exercise caution here.

October Begin harvesting and sales of fish.

November Continue harvesting. Drain the ponds for overwintering. Start to overwinter fish. Check for diseases and parasites. Sell fish.

December All fish that are to be overwintered should now be in the wintering ponds. Summer growing ponds should be empty and drying out. Sales of remaining fish to continue from holding facilities.

18 Carp farming in tropical areas

Since the end of the Second World War there has been a manifold increase in fish farming production throughout the world. In the Northern Hemisphere trout and salmon farming has made great progress, whereas in more temperate climates and in tropical countries, carp and tilapia farming is developing very rapidly, with the help of large government grants. There are however exceptions to this broad outline statement, for instance, in Siberia, where winter temperatures sometimes fall to – 50°C, intensive carp farming on a large scale is taking place in the warm water from power stations. In complete contrast – trout farming is taking place in Kenya in the higher regions of Mount Kenya, where water temperatures never rise above 10–12°C. Even Tenerife, Canary Islands, has a flourishing carp farm in the Mount Teide region, which is competing favourably, price-wise, with the sea fish in this area.

For a number of years the author has been involved with the exporting and growing of one species, *Cyprinus carpio* L, in Nigeria. In chapter 5 it is explained how the feeding and growth of this species slows down as lower temperatures appear at the beginning of the winter period. As water temperatures fall to less than 8°C the carp prepares to hibernate. In Nigeria water temperatures in ponds rarely fall below 15°C, and in summer, temperatures seldom exceed

30°C. *Cyprinus carpio* L is tolerant of these extremes of temperature and therefore will feed and grow all the year round. Such carp will reach a weight of 0.7–0.9 kg (1½ lb–2 lb) in one year. Yields of 2 tonnes per hectare have been achieved without the aid of any additional feeding. However there is also a slight drawback in that because of high water temperatures the carp becomes prematurely mature and breeding starts at the end of the first year. It then follows that within four to five generations of breeding the carp degenerate and the offspring will not reach the same weight as newly imported stock from a more moderate climate. This can of course be overcome quite successfully by importing new breeding stock at regular intervals.

Various systems of growing carp in tropical conditions

Perhaps the most suitable areas in which to develop carp farming are those where farming of some form already exists, this is chiefly because farming in tropical conditions cannot take place without an adequate water supply. Furthermore, where farming does take place, there will be a certain amount of waste production, which will be either of animal or vegetable origin. Water is usually in abundance in the rainy periods. Where reservoirs are already in use, carp can be grown in them. Waste products of any kind can be used for supplementary feeding. For example, water enriched with faeces will promote the growth of zooplankton or if used for irrigation purposes, it acts as a fertiliser.

- Intensive carp farming may be possible in streams which carry a constant supply of water, see *Intensive carp farming in warm water.*
- Where farming of any description takes place in the vicinity of lakes, carp farming may be possible using floating cages. Growth in cages can be substantially increased by giving supplementary feeds either by hand, by demand feeders, or by fully automatic feeding systems.

- It is possible to grow carp in rice fields, and in the last few years some Eastern and Far Eastern countries have concentrated on this technique. The exploitation of this method of carp farming gives perhaps the greatest potential of all developments as it represents a complete recycling of all waste material. Vegetable waste from the rice fields is used as a direct food for the carp and encourages zooplankton growth. Faeces from the fish in turn supplies fertiliser to the rice plants.

Polyculture In the last few years experiments have been made to mix other species of warm water fish with carp in an attempt to achieve even higher yields, in a similar manner to that practised in Northern Europe with carp and tench. Fish chosen for these experiments are mainly catfish and tilapia. Many other species are also being considered, and these experiments are still in progress. However, all the present signs show tilapia to be the most suitable fish for polyculture.

Tilapia Both carp and tilapia were amongst the earliest freshwater fish to be cultured. It is generally accepted that some species of tilapia were already being cultivated about 2000 years ago.

The fish belongs to the cichlid family, a familiar name to aquarists all over the world, it is in this family that the most colourful of all freshwater fish are to be found. Unfortunately they are not suitable to be stocked communally in aquariums as they have a tendency to fight particularly when in breeding condition. Tilapia is of African origin and nearly 100 species can be found in all regions of that continent; some in brackish water and even sea water.

Until about 20 years ago, it was generally considered that tilapia were mouthbreeders; *ie* after the mating pair have cleared a part of the river or pond bed by fanning, the eggs are laid, fertilised and then picked up (usually by the female) and incubated in her mouth. In some species this duty is performed by the male. After 12–15 days the eggs hatch and the young fry are kept in the mouth until the yolk sac is absorbed. The fry are then released, but at the slightest sign of danger will return to the mouth until they reach a size which makes this impossible. Scientists have found that the mouthbreeding habit does not apply to all tilapia species. Some species clear a breeding space, spawning takes place, and the eggs are guarded and fanned until they hatch; the parents regularly patrol the breeding area during this period to ensure the safety of the newly hatched fish. These and other characteristics are the reasons behind the decision to split this species into two groups: the mouth breeders were given the name *Sarotherodon* species and the other retained the name *Tilapia*.

Further differences are that the *Sarotherodon* species are mainly plankton feeders whereas *Tilapia* are mainly herbivorous. From the feeding habits alone it is easy to recognise the benefits which polyculture of these species may bring. Obviously the stocking density of carp has to be reduced in such cases, but the overall yield at harvest will be greatly increased. This is due mainly to the fact that all three species have different feeding habits, and yet economically they are a very good combination, as each converts vegetable matter into animal protein. Also, for *Tilapia* and *Sarotherodon* species there is an existing export market to many cities of the world, and in luxury restaurants they are regarded as gourmet food. Like carp, tilapia have the reputation of being excellent food for people suffering heart and circulation problems.

On the negative side, these fish have a limited growth and reach an average of 0.5 kg (approx. 1 lb) in weight, they are fully scaled and somewhat difficult to clean. Also, under pond culture conditions there is a tendency to overbreed; experiments to control this are still going on and some successes have been recorded, particularly with hormone treatment. More information in this field is available from Universities which specialise in aquaculture. In the UK, Aston and Stirling Universities are well known for their expertise in this field.

Carp predators in the tropics

The protection of fish against predators is a problem everywhere, but it is a major difficulty in tropical countries. Birds are the main culprits as they are capable of flying long distances, using thermals to give them extra lift, whilst carrying heavy loads of fish to feed their hungry young. Pelicans, eagles, herons, cormorants and kingfishers are amongst the most troublesome. Whatever the local problem is, the prospective carp farmer should ensure that the right deterrent is on hand to maintain control. Possible measures include the placing of nets over fry ponds, creating fences in the shallow parts of ponds in an attempt to prevent waders entering the water, or by using electronic equipment which is now available.

Diseases and parasites

Diseases and parasites cause problems the world over, but develop very much quicker in tropical regions. Fish in ponds, where there is little or no movement of the water, are at great risk, particularly from ectoparasites (those which attack the fish externally). Treatment for ectoparasites is described earlier in this book.

Fish which are farmed in cages or troughs are much less likely to suffer the effects of these diseases since the water flow ensures that the habitat is less favourable to the organisms in their embryonic stage.

Endoparasites

Endoparasites attack the fish internally; they mostly have the appearance of worms and are less common in the UK than in some parts of Europe where they have been the cause of heavy losses in the past. These ‘worms’ (some species grow up to 40 cm or 16 in) lay their eggs inside the fish. The eggs are eventually expelled with the faeces into the pond, and are so minute that zooplankton, such as *Tubifex*, *Daphnia* and cyclops feed on them. These in turn are eaten by the fish and so the cycle starts again. Waterfowl carry the eggs from one pond to another thereby spreading the disease.

Treatment is possible, using medication prescribed by a veterinary surgeon. The standard method is to add the medication to the fish food.

After harvesting infected ponds, the pond should be completely dried out before liming with 1 tonne of lime per hectare (400 kg per acre). Attacks by larger parasites, such as fish lice and particularly leeches are very common; these owe their wide distribution to the movements of the many species of water fowl endemic to the tropics. Treatment for these parasites is given earlier in this book.

Fortunately bacteriological and viral infections on carp farms in the tropics are quite rare mainly due to the fact that there is no hibernation period for the fish. Hibernation tends to weaken the carp by lowering its resistance to disease in the period immediately following dormancy. Where hibernation does not take place and there is a plentiful food supply the carp are likely to be stronger and more resistant to disease.

19 Koi

Interest in water gardens is growing all the time in the UK, and it is therefore not surprising that pet shops and garden centres are competing for their share of what appears to be a very lucrative trade. Lucrative it may be, but as in most other fields, there is no lasting success without substantial effort.

The most familiar ornamental garden pond fish is the humble goldfish. It is not directly related to the koi, but is the same species as the crucian carp (*Carassius carassius*). The red and golden colours have been achieved by a system of selective breeding, just as the koi was developed from the wild carp. In nature, the crucian carp can grow to 45 cm (18 in) in length. The goldfish reaches sexual maturity when it has grown to a size of 15 cm (6 in) in length. For this reason it is widely regarded, together with its cousin the shubunkin, as a most suitable fish for small and medium sized garden ponds. The goldfish is in many ways far hardier than the koi and can comfortably winter in ponds which are 60–75 cm (2–2½ ft) deep. In other words, the goldfish is the ideal fish for the beginner.

As the new pond keeper progresses and gains in experience, he will eventually seek to raise his standards, and the ultimate in garden pond fish to keep and grow is the koi. The Japanese can be said to be the masters in producing and rearing koi. In Japan the

Fig 87 A British-bred koi

Fig 88 Four-month-old koi

industry has reached such proportions that some specimens are said to be priceless, but then, the Japanese possess all the advantages for this kind of success – the correct water conditions, a favourable climate and the expertise – which is passed on from generation to generation. To import Japanese koi into the UK or other Western countries is very expensive and there is a degree of risk to the fish due to the very long journeys involved. It it not surprising, then, that enthusiasts in Western countries have tried to produce koi. At the present time, koi are successfully produced in several countries including Italy, Germany, Israel, USA, Singapore and also England.

For the last seven years the author has been directly involved in the breeding, rearing and selling of koi. This began on an experimental basis from an imported stock of Japanese fish. The koi were first quarantined for about a month and then transferred to a small growing pond. When the experiment started in the spring, the koi were an average size of 46 cm (18 in) and by the end of the summer season in the pond, some were showing signs of maturity. The promising fish were taken out of the pond and into the hatchery, where the temperature was gradually increased to 23°C. The fish adapted quickly to the new conditions and were feeding well. By the end of January they were ready for spawning. From 8 fish (4 males and 4 females) only one pair was successful, but from these about 150,000 eggs were produced and incubated in Zuger jars. 35% of these eggs were found to be infertile. The hatched fry were fed for 14 days on *Artemia salina* (brine shrimp) and then transferred into growing troughs, where they were given a diet of dried food (at this time a special feed had been developed by Avos Ltd, a Swedish company of fish food manufacturers; but the author believes that this feed is no longer available). The koi were kept for 3 weeks entirely on this food. After this time the diet was changed to trout starter crumbs, and gradually upgraded as the fish grew. Being a very mild spring, it was possible to place the fish in an outdoor pond by mid-April. The pond in question was sheltered and had an abundant supply of zooplankton. By that time the fish numbered approximately 35,000 and had an average length of 4 cm (1½ in). Initially the colours were disappointing, but after 1 month, not only had the fish doubled in size, but had improved quite dramatically in colour. The breeding pair were sanka (two tone shaded), and some 15–20% offspring were about the same colours as their parents. An equal number had very little colour and could be described as very light coloured common carp. A similar percentage were of an over-all reddish colour, and the remainder were an odd mixture of colours.

This first experiment proved a number of facts which are still valid. Firstly, koi can be bred in the UK in commercially viable quantities but breeding must take place in a hatchery with the help of pituitary injections. Basically koi are as easy to breed as any other carp in natural conditions, and many koi owners must have on occasions been delighted when theirs spawned in the garden pond. The problem here is that the natural spawning time is late May or early June, which for commercial purposes is too late (the selling of ornamental fish begins at Easter and ends in August, just prior to the peak holiday season). By far the most popular saleable size is the 7.5–10 cm (3–4 in) fish, therefore the naturally bred koi would not reach this size in the selling season. They would not even reach a

Fig 89 Eight-month-old koi

Fig 90 A 16-month-old koi

reasonable size for overwintering, as their growth rate is slower than that of the common carp. In Japan and other countries with a similar climate, the growing season is much longer and therefore natural breeding presents no problems.

Secondly, zooplankton plays an important role in the colour of koi. This is not to say that a koi, which initially shows little or no colour, will become colourful. What it does mean is, that the colour markings, which are already present, may become more pronounced. The pigment in the zooplankton is the ingredient which enhances the colours of the fish. This is of course not a new discovery; it is well known that flamingos derive their delicate colour from the crustaceans on which they feed. Also the salmon develops a distinctive flesh colour from feeding on crustaceans. Even table carp, when fed on a wheat supplement, become more golden in colour than those fed on barley.

Both salmon and trout pellets are available with added pigmentation. Experiments at Newhay have shown that the use of these improves colouring on koi, but in time, this food colours the water. A number of pond pellet and feeding stick products produce good results, but such foods are too expensive to be used commercially.

Finally, the demand for British produced koi was found to be surprisingly high, even for the first experimental batch; and this demand has not changed up to the time of writing. Whilst it is true that in terms of coloration the Japanese fish is of a very high standard, British produced koi competes favourably with that produced elsewhere. However, so far as health is concerned, the British koi should have a definite advantage as it is not subjected to such high stress levels as imported koi. That is not to say that British bred koi are guaranteed to be free from diseases and parasites, these appear everywhere from time to time, regardless of where the fish has been bred.

20 Comparisons between carp and koi

Breeding As mentioned earlier, it is possible to produce koi commercially in Hofer and Dubish ponds (see *Breeding in Dubish ponds*) but these fish do not reach marketable size in time for the early selling season, which starts at Easter, the most popular size being the 7.5–10 cm (3–4 in) fish.

The induced breeding of koi is carried out exactly as for carp (see *Induced breeding*). It is advisable to have as many breeding fish as possible prepared for the early attempts, as a large percentage of fish are infertile (approx. 25–30%). The possible reasons for this are, that the brood fish may have been treated with hormones, may have suffered permanent stress damage, or could be products of in-breeding. The pituitary material is the same as that used in carp breeding.

To improve colour quality, it is essential to keep accurate records which clearly show the purposes for which each fish has been used. Failure to do this will result in little or no progress in the markings and colours of the offspring. Identification of the breeding fish itself can take place in a number of ways: by taking photographs of the fish; by snipping corners of the tails in different ways (see *Fig 91*). This will not hurt the fish, but it is advisable to apply a touch of acriflavine to the affected part. A small amputation will grow out again in time, but will be slightly deformed

Fig 91 Corners can be snipped off the tail for identification purposes

thus giving an identification for years to come. Another method of identifying fish is by tagging, and this method is favoured by water authorities.

A further advantage of fish identification is illustrated in the following example: a matched pair produced a good number of offspring, but the markings and colours of the offspring proved disappointing. A year later the female was paired with another male, which was similar in markings and colour to the previous one, but the result of this match was noticeably better (see *Ghost koi*).

Carp and koi farming

Koi farming can be very profitable but requires far greater experience and is much more time consuming. In many ways, koi have lost their self supporting instincts; they have become pets. The breeding procedures of koi are exactly the same as for carp. To breed koi, a larger brood stock will be needed in the early stages. These fish must be kept in well prepared ponds and in the best conditions in accordance with their high value. It is advisable to keep and grow koi as near to the keeper's living accommodation as possible, as they are more at risk from predators than carp. They are easily visible in the water compared to the camouflaged carp, which tends to keep to the bottom of the pond. Koi require a constant temperature, and ponds must be protected from cooling winds. There are a number of ways to do this: erecting wind breaks, plastic tunnels, or even solar heating. At Newhay a unique method was developed. The Newhay farm is situated alongside the River Ouse on a large U bend of the river. As the river is tidal there is a regular flow of driftwood, which is deposited directly opposite the farm. This driftwood is collected and cut into even pieces. It is then fed into an incinerator, which heats two ponds, each 0.1 ha (approx. $\frac{1}{5}$ of acre) in size. By heating only one pond at a time it was possible to raise the temperature of the water by 4°C if conditions were favourable (little or no wind, and the air temperature not less than average). In windy conditions the water temperature would not rise by more than 1–1.5°C, but as the fish sense the heat source, they gather around the heating pipes at the bottom of the pond, and by doing so are able to survive a sharp drop in temperature, as usually experienced in early April. By heating the pond in this way it is possible to have koi of 7.5 cm size – which were born in February of that year – available to sell in May. As the selling season for koi normally starts at Easter, and the fish bred as described above are not ready for sale until May, overwintered carp can be offered for sale from Easter onwards, thereby bridging the gap in the market.

Of course not every carp farmer is fortunate enough to have a supply of driftwood brought to his door, but the object of this example is to alert the koi farmer to consider his own methods of heating, and to encourage him to find other sources of cheap fuel, such as straw, sawdust, bark waste from timber yards or furniture factories, waste oil, or other burning material.

Koi and carp – can they be grown together?

As these fish are the same species they can be grown together, but it is not advisable for the following reasons: carp grows faster than koi, therefore in the growing pond the koi would be at a disadvantage. Carp are more resistant to disease and can be carriers. In other words – carp are robust, and the koi already at a competitive disadvantage, are less resistant to disease. Furthermore, separation of the species when taking them out of the pond presents difficulties. All fish suffer stress when netted or trapped, and the procedure of separating these two varieties, will increase that stress.

For koi farmers who do not produce their own fish there is a disadvantage that domestic fish cannot be imported, whereas ornamental species, including koi, can. It is therefore most unwise to mix domestic fish with imported koi, simply because the home-bred fish should be free from disease, as fish farms are controlled to some extent by the Ministry of Agriculture, Fisheries and Food and also by water authorities. Whereas imported koi are not subject to these restrictions and may well be a source of infection to the native fish. If the two are put together, it is not difficult to imagine the possible consequences. For example: if a fishing club buys a quantity of stock from such a pond, disease could be transferred very easily to other water areas. If koi are not mixed with domestic fish any disease, which may be present, is unlikely to spread beyond the garden pond. It is for this reason that the situation is constantly monitored by MAFF, and it would not be surprising, to see some changes in the existing regulations in the foreseeable future.

21 Ghost koi

The ghost koi is the result of a cross between a female mirror carp and a sanko koi male. Production was first attempted at Newhay in the early 1980's and the result was very interesting. None of the young fish had the appearance of the mirror carp, there was a rapid growth rate and about 50% had the appearance of common carp. The other 50% had shiny silvery scales and when viewed from above, the head had a skull-shaped marking. Although the spawning male had dual colouring of white and blue, none of the offspring showed any trace of the distinctive blue markings at

Fig 92 A ghost koi

first, but as they grew larger, bluish rings around the scales appeared on some fish, giving these individuals a very striking appearance (see *Fig 92*). This experiment was repeated a number of times until the male was accidentally lost. Another male of similar appearance was then used, but the results were disappointing – all the offspring had the appearance of the common carp. Two further males, again of similar appearance, were tried before encouraging results were again achieved.

Ghost koi are exceptionally vigorous – they grow well and overwinter well, and it is not surprising that they are very popular.

22 How to start koi farming

Geographically the Humber–Mersey line must be regarded as being the most northern part of the UK where koi can be reared in open ponds, and then only in selected and secluded places. It is not necessary for koi ponds to be as large as those required for the rearing of table carp, as the bulk of the sales are of smaller sized fish. A pond of 0.2 ha size with a filter aeration unit installed easily supports 30,000 koi of up to the 7.5 cm (3 in) selling size, but of course, more than one pond is needed. The minimum requirement is for three further ponds: one for wintering, one for breeding and one growing-on pond. These should be developed as for domestic carp operations (see section on building a carp pond). The koi ponds should be limed and if only the rearing of koi is intended, particular attention should be given to the quarantine and disinfecting facilities. If it is intended to produce koi it is advisable to try rearing domestic carp in the first instance. This will not produce the same profit margin, but they are hardier than koi and represent a good way of gaining the necessary farming experience.

Trapping and netting koi

Koi selling is a seasonal trade and the smaller sized fish are the most popular. Therefore it is necessary to take small fish from the ponds regularly during the growing season. This operation should be approached with patience and care and planned in such a way that fish are held in tanks for the least possible time. There are specialised traps available, made in a variety of mesh sizes. For small koi a trap with very small mesh must be used to ensure that the head of the koi cannot pass through the mesh causing injury to the gills.

It is a relatively simple matter to make a trap for small koi using plastic material. Such a trap is not only extremely efficient – it will catch from both sides – but is also inexpensive to make (see *Fig 93*). Traps should be laid in the evening, with a few pellets thrown over them as bait, they can then be emptied the following morning by removing the lid on top of the trap. The koi should be very gently shaken into a bucket filled with pond water before being transferred into a tank. Care should be taken to ensure that traps are only emptied in suitable temperatures. Both air and water temperatures should not vary from cold to warm by more than 2°C, (36°F) or from warm to cold by more than 1°C (34°F).

Koi, particularly young fish, are vulnerable to chill, and in severe cases this can result in their loss. The best time to lift the traps is in the morning when temperatures are most likely to be suitable. The same conditions apply (suitable temperatures and mesh sizes) when ponds are netted to catch large quantities of koi. Netting must be carried out very slowly to avoid

Fig 93 DIY fry trap

disturbing the bottom of the pond. Such disturbance blocks the net, making it heavier and causing even more disturbance; the resulting residue enters the gills of the fish, causing discomfort and possible injury. Fish should be removed from the net as swiftly as possible and placed in a prepared container with good, clean, well oxygenated water.

Koi for sale

Ornamental fish keeping has grown in popularity in many countries, including the UK, and this is understandable as we become more and more urbanised. Natural beauty spots are swallowed up and people attempt to compensate for this by recreating a piece of living nature in their own gardens. Fish keeping for pleasure has been with us for a number of years, and if done properly is very rewarding and greatly enhances any garden. This hobby has many attractions; to some it is a challenge, to others a relaxing pastime, whilst to children it is both fun and educational.

It is not surprising that in recent years pet shop owners and garden centres have put more emphasis on ornamental fish than ever before. It is at such outlets that koi, along with other ornamental species, are offered for sale. There are also a number of establishments which now specialise in koi alone, but that is not to say that good quality koi can not be bought at non-specialist places. The customer, of course, will make his own decision, but it is advisable, to refrain from buying a koi simply because of its particular markings. The first consideration should be one of health – the best coloured fish is of no value if it is not a healthy fish. For the novice it is always advisable to have an experienced person present when buying koi.

Koi in the garden pond

A good sized garden pond will provide a haven of peace for the koi when it finally arrives after all the stresses suffered as a result of netting, transporting, fluctuations in water temperatures and water quality. There are some enthusiasts who like to keep koi in cold water aquariums, but this is not advisable: a fully grown koi will be something in the region of 60 cm (24 in) long but if not provided with adequate living space it will become stunted. Another disadvantage is that the koi has retained its bottom feeding habit and will therefore stir up the sediment in the aquarium, clouding the water. Even small garden ponds are not really suitable for koi, and are better stocked with goldfish, shubunkins, golden orfe, golden tench etc. It is not just the size of the pond which has to be considered when keeping koi, depth and volume of water are equally important; small ponds heat up and cool down too quickly and these conditions are not acceptable to the koi which may respond by attempting to jump out of the water.

The cost of building a koi pond can be kept to a minimum and can be easily undertaken by the handyman. It can be constructed from concrete and coated with a good sealing compound. Pond liners of all sizes and qualities are available, also a wide range of fibreglass ponds, some in very interesting shapes. The koi pond should be at least 1 m (3 ft) deep (preferably 1.15 m or 3 ft 6 in) at one end and should hold 4,500 *l* (approximately 1,000 gal) of water. As it should be a feature of the garden, the position and shape of the pond is very important. For example, an unconventional looking pond will scarcely fit into the garden of a Georgian property.

A pond with a large water surface will develop algae in summer and the water will quickly become green. This is not harmful to fish, but unfortunately, they are not visible through it. This can be partially prevented by increasing the number of water plants growing in the pond. Also chemicals are available to control this condition, but by far the best method is to install a filter unit. The cost of this is governed chiefly by the price of the submersible pump, as these are rather expensive. The filter unit can also be bought complete, but it is not difficult to make one.

Plastic storage tanks are very suitable for filtration purposes, and for a 5,000 *l* (1,000 gal) pond a 250 *l* (50 gal) sized tank is about the right capacity. *Figure 94* shows how the filter tank should be prepared; the water supply pipe from the pump goes through the tank and then connects up to a plastic pipe, which has previously been drilled with 3 mm (⅛ in) holes every 12 mm (½ in). On the bottom are two pipes of the same size drilled with holes as previously described. On the top of these lower pipes is a layer of plastic mesh, which is held in place with bricks. Washed pea gravel is then placed on top of the the plastic mesh, and the tank filled to three quarters full with the gravel. A layer of porous foam 12 mm (½ in) thick should be placed on top of the gravel. The filter works as follows: the submersible pump at the deepest end of the pond is switched on, the water is pumped into the filter which then distributes the pond water evenly through the drilled holes. The water passes through the foam, where it deposits coarse sediment before passing through the gravel. The filtered water then

Fig 94 Pond filter

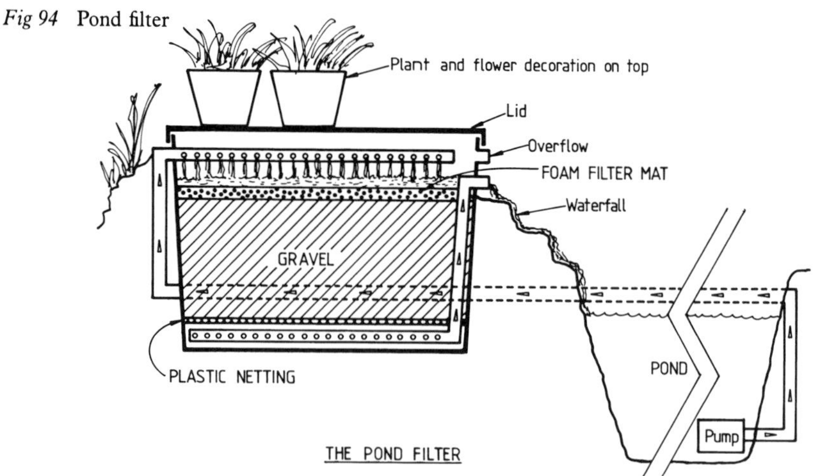

passes through the holes of the lower pipes, and, through its own pressure will raise and discharge, thus forming a waterfall before returning to the pond as clean recycled water. An overflow pipe should be installed above the two outlet pipes in case the filter should become blocked for some reason; this acts as a safety precaution to prevent the pond from being pumped empty. It is not necessary to run the filter continuously, timing depends on the stocking density of the pond and the efficiency of the filter. The filter can be buried by using some of the surplus soil excavated from the pond. The top of the filter should be covered by a lid which can be ornamental. Every two weeks or so, the lid should be opened to check and, if necessary, clean the foam. This also applies to the small filter at the end of the submersible pump. The pond should then be decorated with one or two water lilies, and the surrounding area can be made into a rock garden, incorporating a waterfall (see *Figs 95* and *96*). Through the waterfall action, the water will become oxygenated and it is then unnecessary to have any oxygenating plants. Some marginal plants can be used to enhance the shallow end of the pond.

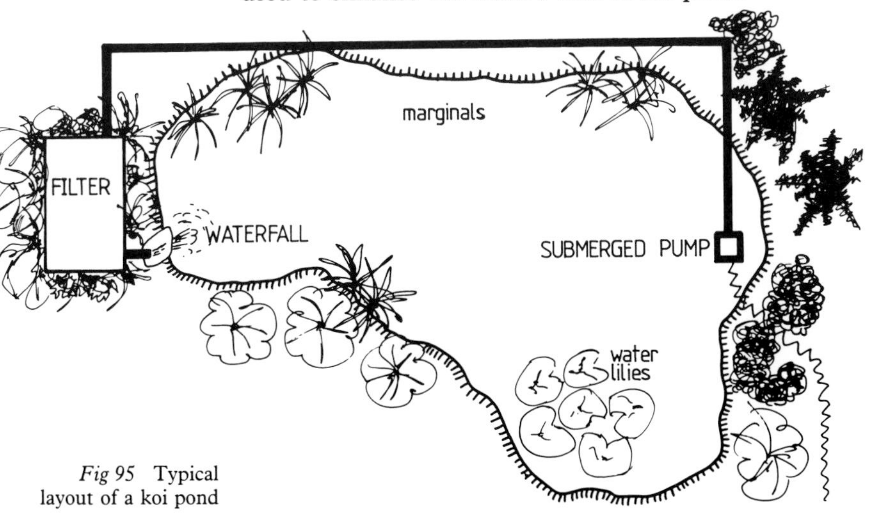

Fig 95 Typical layout of a koi pond

Fig 96 Small koi pond in early spring

The koi will soon adapt to its new home, and if fed regularly with floating pellets or sticks at the same place and by the same person, they become very tame and can be trained to take food from the hand. They will even allow a gentle touch. A well stocked and decorated koi pond is truly beautiful and well worth the effort of making.

Parasites and diseases

Most common carp parasites have already been mentioned and treatment for koi is exactly the same. If treatment should be necessary it is better to use a concentrated bath solution than to attempt treatment in the pond. The fish can be easily taken out of the pond and stored for a time in suitable containers, such as plastic bins, large aquaria or even the household bath whilst the garden pond is disinfected and refilled. It is quite safe to return koi to a pond which has been freshly filled with tap water, provided the water temperature and pH are the same as that of the storage container (tap water usually has a pH value of 7.0). After the parasite or disease has been identified, the following check list should be referred to for information on how to treat the condition.

Disease or parasite	*Treatment*	*Dose*	*Time*
Leeches	Lime or cooking salt	2 mg/*l* water 25 mg/*l* water	5 seconds 10 minutes
Costia *Ichthyophthirius*	Malachite green	0.1 g/m^3 water	Treatment in pond. Repeat after 1 week if necessary
Costia, Dactylogyrus *Gyrodactylus*	Cooking salt	25 mg/*l* water	10 minutes
Carp louse *Costia*	Potassium permanganate	1 g/10 *l* water	10 minutes
Costia, Dactylogyrus *Gyrodactylus*	Formalin	20 ml/*l* water	30 minutes

If koi die for no apparent reason, a water test should be carried out. If this is satisfactory and no parasites or diseases detailed in this book can be identified, it is possible that the fish is suffering from a viral infection. As koi, and especially adults, are usually very vulnerable, the advice of a veterinary surgeon, the water authority or a firm specialising in fish diseases, should be sought.

Treatment for virus diseases can be given using antibiotics which are mixed into the food, but this method will only be effective if the fish are still feeding. Bathing in a solution of antibiotics, or injecting the drugs, may save the fish. Antibiotics can only be obtained on prescription from a veterinary surgeon, and not all are freely available in the UK, such as Prefuran or Furanace, which have to be imported.

Killing a koi

Knowledge of fish diseases increases with the passing of time. Treatments are usually effective in cases where a correct diagnosis is made. In garden ponds some diseases, unfortunately appear so suddenly that the affected fish are in a state of advanced infection before the condition is apparent, by which time it may be too late for treatment; in such cases the fish must be killed. To kill any creature, for some people, demands a great deal of courage, and sometimes, in the haste to get it over with, the keeper will use an inhumane method, such as flushing the fish down the toilet. The fish is quickly out of sight, but its suffering is by no means over. However unpleasant it may seem, the quickest and most humane way is to take the fish from the pond, hold it with a cloth to prevent it from jumping, and if the fish is small, cut the head off with a sharp knife, just behind the gills. Scissors can also be used. Large fish should be stunned by administering a sharp blow to the head, and then cut through the spine behind the gills. A dying fish should not under any circumstances be released into an open water in the misguided belief that this action will give it a chance to live. This can result in the fish infecting open waters and thereby endangering other fish; it is both very irresponsible and illegal.

Feeding koi

Feeding koi commercially in large ponds is the same as feeding domestic fish. It is advisable not to be hasty by starting additional feeding with the first few warm days in spring. Koi just coming out of hibernation look around for bits of algae and plants to clean their alimentary system. Giving high protein food too early has no advantage as the fish has difficulties in digesting the food, which also clouds the water. Koi only really start feeding and digesting properly, when the temperature reaches 12–15°C (54–59°F) only then should any additional feeding commence, and small quantities only.

Feeding koi in the garden pond

Even in the garden pond insects enrich the koi's diet. A few water beetles will find their way into the pond, and insects which accidentally fall into the water, will be consumed by the koi before they are able to take off again. Mosquitoes lay eggs in ponds, some of which develop to the larvae stage. All of these are very welcome food for the koi, but insufficient for its needs.

The appetite of the fish increases with the rising temperatures, and at this stage supplementary feeds must be used. With the development of floating pellets and the floating stick, koi feeding has become simpler with more certainty of results. It is best to offer food in small quantities, frequently, than to give too much food at any one time. The koi, along with all carp, has almost no stomach, and because of this, the species is quite unable to digest food in the same way as most other animals. Food taken in by the koi stays in the digestive system for no more than 5 hours before it is expelled, and this means, that they can be fed every 5–6 hours. However, in practice once a day is sufficient as there is a certain amount of natural food in the pond, as already mentioned. Koi feed best when the water is at its warmest, and this is in the afternoon or evening, daylight is not important as the fish relies more on smell than sight. It is important to note that feeding should cease when the water temperature rises above 25°C (77°F) or falls below 10°C (50°F). Food must be of good quality and fresh. The emphasis here is on fresh, as koi food has a limited shelf life and after a given time, the essential vitamins will be lost. Good koi food is vacuum sealed giving a longer shelf life; however, once opened it should be used within the specified time. Thus the koi keeper should be wary of bulk buying for the reasons given.

Koi breeding for the hobbyist

Female koi reach sexual maturity at 4 years, and males at 3 years. The koi hobbyist will often optimistically observe his first spawning on a spring morning in the hope that he will be rewarded with large quantities of baby fish. This early anticipation frequently ends in disappointment when few, if any, juveniles result. The main reason for lack of success, is due to the parent fish (and other fish in the pond) eating the eggs and any fry which hatch. There will be greater recruitment success where the pond is overgrown with oxygenating plants, such as *Elodea densa* or *E. crispa*. Another reason for failure is that the eggs may become covered in sediment, (stirred up by the intense activity of the breeding fish) thus depriving them of oxygen.

The following spawning procedure using artificial 'plants' has been successfully used by the author, and is recommended to amateurs who may wish to try it.

As with other carp species, breeding usually starts in late May or early June. The water temperature must be constant and in excess of 18°C (65°F). All oxygenating plants should be removed from the pond, but water lilies can remain. A plastic spawning medium should be prepared: a plastic table cloth or similar plastic sheet should be cut into strips 6 mm (¼ in) wide and 20 cm (8 in) long. A dozen of these should be bundled together and a weight fastened on one end. When sunk into the pond this will have the appearance of a bunch of grass. The weight should not be made of lead, and as koi are capable of recognising colour, the material should be green. 20–30 such bundles will be needed.

In the meantime, a suitable container should be prepared – a 1–1.3 m (3–4 ft) aquarium is ideal. The object is to make the koi spawn onto the plastic plants, which are then transferred with the eggs attached, to the tank. As spawning time approaches, the pond must be kept as clean as possible, and special attention must be given to removing the sediment from the corners and from the gravel on top of the water lily pots. The behaviour of the fish should now be carefully monitored for signs of males chasing a particular female, *ie* one which is ready to ovulate. The chasing begins a day or two before spawning takes place, and this is the time to place the artificial plants into the shallow end of the pond. Providing that all oxygenating plants have been removed, the koi will spawn amongst the artificial ones. Spawning usually takes place early in the morning and is completed by 10 or 11 am. Half an hour after spawning activity has slowed down, the artificial plants should be removed from the pond and placed in a clean bucket, which has been filled with pond water, before being transferred to the prepared tank. This will have been filled with tap water two days previously, well aerated and brought up to a temperature equal to that of the pond water (using an aquarium heater and thermostat). Before the plants are taken out of the bucket, they should be gently agitated in the water, to remove any dust and sediment. They should then be placed in the tank for the incubation period. The aerator or filter should be operated at a gentle speed to give the eggs maximum oxygen. At a temperature of 23°C (73°F) for 3–4 days the eggs will begin to hatch. At a temperature of 20°C (68°F) hatching will take about two days longer. The newly hatched koi will attach themselves onto the plants or to the side of the container, and after 3–4 days (or 4–5 days at lower temperatures) the larvae will be free swimming, after filling their swim bladder with atmospheric air. They will now be looking for food, their initial supply from their yolk sac being exhausted. It is at this stage that all the artificial plants are taken out, carefully dried and stored away for future use.

The advantage of using artificial plants is that they are so much easier to clean, they do not decompose, and do not harbour micro-organisms which could have an adverse effect on eggs and fry.

The fry must now be fed and unless the hobbyist is prepared to hatch *Artemia salina* as described in the chapter on induced breeding, the next best food for the fry is hard boiled egg yolk. The disadvantage in using egg yolk is that it fouls the water very quickly, making regular cleaning of the tank necessary. This feeding with egg should continue for a period of one week, and then the diet should be changed to newly hatched *Daphnia*, which are in abundance at this time of the year. Armed with a zooplankton net and a plastic bag or bucket, a visit to almost any village pond is likely to yield a catch of *Daphnia*, but as they move around in the pond, several attempts may well have to be made before they are located. Before feeding to the fry these crustaceans should be graded through a fine hand net and only the smaller ones used as fry food. The large *Daphnia* will provide a welcome addition to the diet of the adult koi, but it should be remembered, that the filter unit must be turned off when feeding live food.

The small koi can be raised in the aquarium or tank until they reach the size of approximately 2.5 cm (1 in). Ideally they can then be placed in a special growing pond but if this is not possible, perhaps a corner of a pond could be netted round, thus providing a separate area for them, away from the adult fish. When the young koi have grown to about 4 cm (1½ in) in size, it is safe to allow them to mix with other fish in the pond.

Growing *Daphnia* for young koi

Whenever possible, it is preferable to use a natural feed for koi and if a small area of the garden can be spared, it is not difficult to set up a *Daphnia* pond. An old bath or any other container, made from non-corrosive material, buried in the ground (to hold the water temperature more constant) is suitable. It should then be filled with tap water and manured; well rotted cow or sheep waste is suitable, as are grass cuttings. Very good and even cleaner alternatives include household flour, dried blood or liquid plant fertiliser. These three can be liquidised and them drip fed into the pond, and in this way, the flow can be regulated to coincide with natural fluctuations in the *Daphnia* population and to prevent the smells which occur with manuring.

Although the production of *Daphnia* can begin naturally by the introduction of a single egg by a bird or animal, it is better to put in a 'starter' which, if not otherwise available, can be purchased from a water garden centre or pet shop.

23 Summary of the koi keeper's year

January/February The koi should be hibernating and be seen to be almost motionless at the deepest end of the pond. They will only move around during mild spells at this time of the year, in which case, a little food should be offered. If ponds are frozen over, a hole should be made by pouring hot water onto the ice. Under no circumstances should the ice be broken by striking with any object. The resulting shock waves from such an action can easily lead to haemorrhages and subsequent loss of fish. Electric pond heaters can be used to keep a small hole in the ice open without raising the temperature in the pond. Any snow and slush on the pond ice should be removed. The filter unit should have been switched off prior to wintering the pond.

March/April As daylight lengthens, the air and water temperatures will rise. The koi become more active and will look around for food. At this stage it is not necessary to feed the fish as they will first take algae and bits of plants to cleanse their digestive systems. The temperature should be checked with a thermometer and when it rises above 10°C (50°F), a little food should be offered. The pond heater can now be switched off. For a beginner, this is the best time to dig a new pond.

May/June When the air and water temperatures are equal, it is advisable to take the koi out of the pond for a thorough inspection. This should be done patiently and carefully, using a large net. Any parasites and diseases are more easily diagnosed at this time of the year. Detection and treatment of any disease, which may be present, should be carried out in accordance with the instructions given in this book (see *Diseases and parasites*).

This is the time of year for introducing new stock to the pond if desired. However, if there has been any treatment of the pond for disease or parasites, the introduction of new stock should be delayed for at least a month. A careful examination should be made of any new stock before placing in the pond to ensure that they are free from infection. A bag containing the new fish should be floated on the water surface for 30 minutes before the fish are released. The bag must be given a protective cover if it is exposed to strong sunlight and opened in such a manner as to allow the pond water to run slowly in, and then tilted so that the koi can swim out. Regular feeding is now necessary.

July/August This time of year brings the best growth conditions for the fish. Pond water temperatures should average about 20°C (68°F) during these months. The koi should be fed at least twice a day in order to build up strength and reserves for the coming winter. On very hot and thundery days the water temperature in the pond may reach 25°C (77°F). If this should occur, all feeding should be temporarily suspended, as feeding in these conditions could cause digestive problems for the koi, and result in sudden build up of toxic waste in the water; under these conditions the high oxygen requirement of the koi would not be met. A symptom of this is the koi gasping for air on the surface of the water. The condition can be alleviated somewhat by spraying tap water on the surface of the pond and also by increasing the aeration of the water through the filter unit.

September/October During these months the water temperature will gradually decrease and feeding should be reduced accordingly. However, food should be offered for as long as koi will take it. Larger fish will stop feeding when water temperatures fall just below 10°C (50°F), smaller koi will stop feeding at temperatures of 6–8°C (43–46°F). Decaying plants should now be removed from the pond and marginals cut back. The pond should be either covered with a net or kept free of falling leaves. As the temperature continues to fall the filter unit should be switched off.

November/December The koi should be starting to hibernate once again and feeding should cease. Only in exceptionally mild periods should any food be offered. If the pond freezes over, the pond heater should be switched on.

24 Carp as a sport fish

Carp is the ideal fish with which to stock static waters for angling purposes. Gravel and sandpits, brickyard ponds, large irrigation ponds and even village ponds can be suitable. The stocking density, however, should not be nearly as high as in controlled farming ponds (see section on stocking density for the growing season), especially if other species are already present in the pond. As a guideline – for a pond of 0.4 ha (1 acre), containing an unknown small quantity of coarse fish, not overgrown with underwater plants and with no inlet of fresh water, no more than 40–50 C3 size carp (a total weight of approx. 45 kg or 100 lbs) should be introduced. Ponds of the same size, which have an inlet/outlet of fresh water will support about twice this number. There is no advantage in overstocking, for two reasons:

- Where the water is static there is a possibility of a serious oxygen deficiency in the hot summer months.
- Every pond has its biological limits and if exceeded, the fish will simply not grow and in such conditions would become stunted.

Stunted fish have no appeal to the angler. But the situation is, of course, very different where ponds can be regularly emptied and limed – in other words, controlled ponds. Such ponds can support 1,200–2,000 C3 size fish per hectare (500–800 per acre).

Ponds for carp fishing

The rules here are the same as those for carp farming. Basically, a well maintained pond will produce good fish. It can be said that a fishing pond is more difficult to maintain than one for farming, as most of the former cannot be drained. In many cases these ponds are privately owned and are let to angling clubs and syndicates and for day fishing. As far as day fishing is concerned the onus lies with the owner to ensure, that the pond is in a suitable condition and well stocked at all times. Clubs and syndicates often stock their own waters, and if it is decided to introduce carp, the following points should be considered:

– The situation of the pond. Ponds in peaty surrounds are usually acid and a pH test will reveal this; if the pH value is less then 5.0, the pond is unsuitable for carp. Even liming such a pond is unlikely to improve its potential for this fish. To be suitable for carp stocking, ponds should have a pH of 7.0–8.5, but ponds showing slightly lower figures, can be successfully limed (see section on liming the ponds). Ponds surrounded by deciduous trees are also likely to have a low pH value. Leaves which have fallen into the pond over a period of years, will have built up an acid layer at the bottom of the pond, and in time this lowers the pH of the water. Such a situation can be improved by thinning out the trees on the north and west sides of the pond (*ie* those which receive the prevailing winds), or removing them altogether and replacing with conifers. The trees on the other two sides of the pond can be left, as the falling leaves are blown away from the water by the prevailing winds. Permission from the owner is, of course, required before such changes are made.

  Rotting water plants of all types, such as bulrushes, reeds and submerged plants can cause acidification as well as causing an accumulation of ammonia in the pond.

– On the other extreme ponds with a clay or limestone

bottom may have an excess of alkaline in the water. Carp are tolerant of alkalinity from 9.0–10.5, but do not breed under such conditions.

– The hardness of the water has some influence on the well being of the carp. Carp are very tolerant of hard water, and in fact show a marked preference for it. For this reason ponds situated close to tidal rivers are ideal for carp, as water from the river seeps into the pond. Carp in these ponds will thrive as also will the zooplankton.

Holding the catch

Ideally the angler, who has been fortunate enough to land a large carp, examines it, and perhaps takes a photograph, before returning the fish to the water. It is difficult to keep a large fish in, what is probably, an inadequate sized keep net, particularly in warm weather conditions, when the increased activity of the fish could well result in injury. Large fish must never be mixed with smaller ones in the same keep net, the latter are easily damaged, or even killed. Damaged fish are on release, very susceptible to fungus infection, which could rapidly spread through the pond.

Advice for the angler

The angler should have some understanding of the lifecycle and habits of the carp such as that which the author has attempted to cover in this book. Only a good healthy fish is capable of putting up a struggle once the angler has lured it to take the bait. All anglers, who have ever had a big carp on the line will know that it is not the easiest of fish to land, and the older the carp the more difficult it is. Carp are very sensitive and are suspicious of disturbances, both inside and outside the pond. Movement around the banks of the pond creates vibrations, which, however weak, will be detected by the carp. If a carp pond is netted during the feeding season, the large carp remaining in the pond will stop feeding for 2–3 days. For this reason the wise carp farmer will not net his ponds during the feeding season.

Carp are connoisseurs, in that once they acquire a taste for particular food they are not easily persuaded to change to another. On farms where demand feeders are used, it is easy to prove this. For instance: carp can be fed on any type of pellet (trout, rabbit or carp pellet), but if they are offered those which contain exactly the same ingredients, but made by different manufacturers they will notice immediately and react by stopping all feeding for a time. Anglers intending to catch carp are well advised to offer the bait for a day or two before fishing. The bait must be exactly the same each time, otherwise their chances of success in landing a carp will be limited.

The best times for carp fishing are known to be at dawn/early morning, which is their normal feeding time. However, because the carp relies more on smell than on sight, it is possible to fish successfully in total darkness. On dull and overcast days the carp may well feed at all times of the day. On the other hand, during periods of bright sunshine they do not feed and can be observed in a favourite pastime of lazing around in the sunshine; during these periods they show little interest in feeding.

As the carp is a bottom feeder, the bait should be placed on or near to the bottom of the pond. Even large carp like to move about the pond in small groups constantly digging for food. The quiet observer will be rewarded by seeing muddy rings appear on the surface, indicating where this is taking place.

By contrast, there are stories of fly fishermen hooking carp whilst fishing for trout; this is certainly possible as the author has often seen carp surfacing and catching flies. If carp are observed jumping on a summer evening, it is a clear indication of good healthy specimens.

Useful conversions

LENGTH

1 centimetre (cm)	= 0.3937 inches
1 metre (m)	= 1.0936 yards
1 kilometre (km)	= 0.6214 miles
1 inch	= 2.5400 cm
1 mile	= 1.6093 km

AREA

1 sq. metre	= 1.1960 sq. yds
1 hectare (ha)	= 2.4711 acres
1 sq. km	= 0.3861 sq. mile
1 acre	= 4046.9 sq. metre

VOLUME

1 litre	= 0.2200 gallon
1 pint	= 0.5683 litre
1 gallon	= 4.5461 litre

WEIGHT

1 gramme (g)	= 0.0353 oz
1 kilogramme (kg)	= 2.2046 lb
1 tonne (*t*)	= 0.9842 ton
1 ounce	= 28.350 g
1 pound	= 0.4536 kg
1 ton	= 1.0161 tonnes

Index

Books published by
Fishing News Books Ltd

Free catalogue available on request

Fishing boats of the world 3
The fishing cadet's handbook
Fishing ports and markets
Fishing with light
Freezing and irradiation of fish
Freshwater fisheries management
Glossary of UK fishing gear terms
Handbook of trout and salmon diseases
A history of marine fish culture in Europe and
North America
How to make and set nets
Introduction to fishery by-products
The lemon sole
A living from lobsters
The mackerel
Making and managing a trout lake
Managerial effectiveness in fisheries and aquaculture
Marine fisheries ecosystem
Marine pollution and sea life
Marketing in fisheries and aquaculture
Mending of fishing nets
Modern deep sea trawling gear
More Scottish fishing craft and their work
Multilingual dictionary of fish and fish products
Navigation primer for fishermen
Netting materials for fishing gear
Ocean forum
Pair trawling and pair seining
Pelagic and semi-pelagic trawling gear
Penaeid shrimps — their biology and management
Planning of aquaculture development
Refrigeration on fishing vessels
Salmon and trout farming in Norway
Salmon farming handbook
Scallop and queen fisheries in the British Isles
Scallops and the diver-fisherman
Seine fishing
Squid jigging from small boats
Stability and trim of fishing vessels
Study of the sea
Textbook of fish culture
Training fishermen at sea
Trends in fish utilization
Trout farming handbook
Trout farming manual
Tuna fishing with pole and line